Successful Digital Transforma

Other publications by Van Haren Publishing

Van Haren Publishing (VHP) specializes in titles on Best Practices, methods and standards within four domains:

- IT and IT Management
- Architecture (Enterprise and IT)
- Business Management and
- Project Management

Van Haren Publishing is also publishing on behalf of leading organizations and companies: ASLBiSL Foundation, BRMI, CA, Centre Henri Tudor, CATS CM, Gaming Works, IACCM, IAOP, IFDC, Innovation Value Institute, IPMA-NL, ITSqc, NAF, KNVI, PMI-NL, PON, The Open Group, The SOX Institute.

Topics are (per domain):

IT and IT Management
ABC of ICT
ASL®
CMMI®
COBIT®
e-CF
ISO/IEC 20000
ISO/IEC 27001/27002
ISPL
IT4IT®
IT-CMF™
IT Service CMM
ITIL®
MOF
MSF
SABSA
SAF
SIAM™
TRIM
VeriSM™

Enterprise Architecture
ArchiMate®
GEA®
Novius Architectuur Methode
TOGAF®

Project Management
A4-Projectmanagement
DSDM/Atern
ICB / NCB
ISO 21500
MINCE®
M_o_R®
MSP®
P3O®
PMBOK® Guide
Praxis®
PRINCE2®

Business Management
BABOK® Guide
BiSL® and BiSL® Next
BRMBOK™
BTF
CATS CM®
DID®
EFQM
eSCM
IACCM
ISA-95
ISO 9000/9001
OPBOK
SixSigma
SOX
SqEME®

For the latest information on VHP publications, visit our website: www.vanharen.net.

Successful Digital Transformation

A survival guide for managers and executives

Marc Beijen

Van Haren
PUBLISHING

Colofon

Title:	Successful Digital Transformation
Subtitle:	A survival guide for managers and executives
Author:	Marc M. Beijen
Publisher:	Van Haren Publishing, 's-Hertogenbosch – NL, www.vanharen.net.
ISBN Hard copy:	978 94 018 0771 5
ISBN eBook (pdf):	978 94 018 0772 2
ISBN ePUB:	978 94 018 0773 9
Edition:	First edition, first impression, June 2021
Lay out and DTP:	Coco Bookmedia, Amersfoort - NL
Copyright:	© Van Haren Publishing and M.M. Beijen, 2020, 2021

For further information on Van Haren Publishing, e-mail to: info@vanharen.net.

Contents

PREFACE BY ERIK BEULEN VII

PREFACE BY MARCO KEIM IX

RECOMMENDATIONS XI

1 DIGITAL REVOLUTION IN THE 21ST CENTURY 1

1.1 Challenges facing organizations 1
1.2 Assignment for managers and executives 2
1.3 Drivers for change 5

2 DIGITAL TRANSFORMATION IN PERSPECTIVE 9

2.1 Phase model for digital transformation 9
2.2 Not one but two digital transformations? 11
2.3 Five digital breakthroughs 16

3 DIGITAL STRATEGY AS A STARTING POINT 19

3.1 Elements of a digital strategy 19
3.2 Framework for business transformation 23
3.3 Digital breakthroughs projected onto the framework 28

4 DIGITAL BREAKTHROUGH 1: DATA-DRIVEN ORGANIZATION 31

4.1 Why is data so important? 31
4.2 Problems and challenges with data 33
4.3 What does a data-driven organization look like? 34
4.4 How to become a data-driven organization? 37

5 DIGITAL BREAKTHROUGH 2: SMART, DIGITAL PROCESSES 41

5.1 Why are smart, digital processes so important? 41
5.2 Problems and challenges with smart, digital processes 42
5.3 What does it take for smart, digital processes? 44
5.4 How to become a digital, process-oriented organization? 50

6 DIGITAL BREAKTHROUGH 3: BRILLIANT CUSTOMER EXPERIENCE 53

6.1 Why is a brilliant customer experience so important? 53
6.2 Problems and challenges with customer experience 54
6.3 What do you need to achieve a brilliant customer experience? 56
6.4 How do you develop a brilliant customer experience? 61

7 DIGITAL BREAKTHROUGH 4: AGILE AND RESILIENT ORGANIZATION .. 65

7.1 Why is an agile and resilient organization so important? 65
7.2 Problems and challenges in IT processes and controls 67
7.3 What do you need to become an agile and resilient organization? 69
7.4 How to become agile and resilient? 72

8 DIGITAL BREAKTHROUGH 5: DIGITAL BUSINESS MODELS 75

8.1 Why are digital business models so important? 75
8.2 Problems and challenges with digital business models 79
8.3 What do you need to set up a digital business model? 81
8.4 How to develop a digital business model? 84

9 MOVING SUCCESSFULLY FROM DIGITAL STRATEGY TO EXECUTION ... 87

9.1 Digital strategy as part of the change cycle 87
9.2 Prioritizing: an initial ranking 90
9.3 Eight core principles for successful change 92
9.4 Finally: and now? 95

LITERATURE LIST 99

EPILOGUE 103

ABOUT THE AUTHOR 105

Preface by Erik Beulen

Professor of Information Management, Tilburg University; Academic Director Executive Master Information Management, TIAS School for Business and Society

Successful digital transformation is still a major challenge for many organizations. Markets are in a state of flux, technological capabilities and compliance requirements are greater than ever. In these dynamics, organizations must digitally transform. I see insufficiently innovative use of IT among many. After all, it is essential for organizations to continually renew themselves. Executing a strategy once chosen is no longer enough; continuous adoption of innovation is necessary for survival. Emerging technologies such as artificial intelligence, blockchain, internet of things and robotics can be deployed by your organization but also by your competitors. Therefore, focus on adapting the DNA - continuously and radically adapt your organization, processes and ways of working.

This book elaborates on successful digital transformation using a phase model and five digital breakthroughs that can be used in the development of a business strategy to realize ambitions. The phase model presented includes three distinct steps - from product focus in the first phase to customer orientation in the second phase, followed by a third phase in which the customer is at the center. The book describes what is required to take these steps - the checklist that is included provides organizations with direction. This helps in setting up a strategy and developing an effective roadmap for organizations to transform.

However, they will have to ask themselves whether there is enough time to transform sequentially. For many organizations, it is five to twelve and the question is whether there is still time to accelerate by initially focusing only on the no-regret choices. A parallel transformation, skipping the step to the second stage, is sometimes the only option to avoid being pushed out of the market by digital competitors. A risky scenario but sometimes you just must follow through and trust intuition. Leave the drafting of a strategy for what it is.

In addition, the five digital breakthroughs are elaborated. The first, data-driven organization, is the most important. Together with the fourth breakthrough, an agile and resilient organization, these are, in my opinion, the basis for the other three breakthroughs. Much has been said about agile working, although most organizations have not made enough progress in adopting it. Data remains the Achilles heel for many digital transformations - data quality is inadequate, master data management is a concern, and volumes of available data are growing exponentially. So, try to make your organization data driven. By focusing on data, it becomes possible to effectively engage in the introduction of new business models (breakthrough 5), customer experience (breakthrough 3) and improving business processes (breakthrough 2).

Good luck with digital transformation!

Erik Beulen

Preface by Marco Keim

CEO Aegon International and member of the Management Board of Aegon N.V.

It is impossible to imagine today's business without IT. Ten years ago, as an executive, you could say that your focus was limited to the business and the IT department ensured that it was adequately supported; today, this is unthinkable. IT is an integral part of the business for every director, board member or manager. It is no longer a choice; it has become a given! At all levels, from C-level to operations, you need to understand what is going on. As a director, board member or manager, you don't have to be an expert, of course not, but you do have to constantly ask yourself what contribution IT can make to your business and ensure, together with the IT department, that this value is actually achieved. You may now feel a threshold for taking this on. Where should you start, what is important and what is not, as there are so many terms and buzzwords. Where do you get the right information from, how do you make it applicable, who is going to help you?

The author of this book, Marc Beijen, is uniquely positioned to show you the way in this comprehensive matter. I have known Marc for more than 20 years and have enjoyed working with him on a regular basis. From my first years as a young manager, he has shown me the way in, what is for me, the still unknown world of IT and business architecture. Marc's strength is that he not only knows everything about the latest techniques and developments, but that he also translates this into realistic, pragmatic, working solutions and added value for my organization. Marc makes complexity accessible and understandable for non-IT people like me and is, therefore, the ideal author for this book.

Successful Digital Transformation - A survival guide for executives and managers is a valuable reflection on knowledge developed and experience accumulated in daily practice, focusing on the approach to contemporary digital business challenges. Written in understandable language and with a powerful, logical structure, it has become a very accessible book that remains fascinating from beginning to end. It teaches you to understand topics that matter such as the need for digital transformation, the connection with business strategy and the essential differences in starting positions between organizations. What to do next is made concrete and practical by identifying five necessary digital breakthroughs and working them out into clear goals, which should be included in the change cycle as part of your digital strategy.

I read this book with great attention and pleasure. It makes you aware of the need for further digitalization to stay in business. It makes you think about how you can make the difference with modern information technology and gain a decisive competitive advantage. Topics that every manager and executive must deal with and, therefore, a book that every manager and executive should read in my opinion. Take your time to allow all the rich and valuable information to sink in. For example, don't read it all at once but rather chapter by chapter. Let the content sink in and place it in the context of your business before you go any further. This will help you provide the right direction for your business transformation.

Marco Keim

Recommendations

"The book 'Successful Digital Transformation' reads like a modern fairy tale. It's clever how the essence of the field of digital transformation is captured in barely 100 pages. But the theme has also been made very practical, the underlying topics (such as data management, enterprise architecture, change management) have been tied together and explained in a comprehensible way. I think the book fills a great need, because many books and articles on this topic are either full of buzzwords that are not further described, or lapse into academic jargon that is incomprehensible to the average reader. This book sits nicely between academia and practice, without falling into compromises."

– Frank Harmsen - Associate Professor of Digital Transformation, Maastricht University

"I enjoyed reading Marc Beijen's 'survival guide' to digital transformation. To me it read like a digital cookbook! It is not easy to make a good dish from a huge variety of ingredients. A good cook must not only know the flavors and preparation of the ingredients. But also understand which ones go together. And which ones don't. In addition, a good cook knows the required preparation time and thus understands that sometimes he must work on the dessert while also working on the entrée. Marc has managed to help with this complexity. All the ingredients are in there. From technology & data to business models & organization. Including the three different phases in which to work (usually simultaneously). This book thus guarantees a lot of inspiration and cooking pleasure in the digital kitchen; both for the cook himself and for the visitor who expects a delicious three-course menu."

– Sjoerd Blüm, CIO – Schiphol Airport

"Marc Beijen takes a big step with this latest book by adopting the need for digital transformation as a compelling guiding theme. A trigger that comes mainly from outside the organization, not so much from its own needs or capabilities. Alignment is dead! It's all hands-on deck. But then ... how? Marc takes the usually recalcitrant starting situation (organization, skills, legacy systems, etc.) that hinders a

transformation into account in his approach. The digital strategy is transformed into a manageable, integral, transformation calendar. I commend this survival guide to every manager who has to bring such a transformation to a successful end. I also commend this to the supervisors in these ever-faster changing organizations."

– Eric Sluis, who was CIO of Achmea Insurance / holds various supervisory directorships at financial institutions

"Much is written about digital transformation. Very few companies succeed in transforming the business model based on digital technology. With this book Marc Beijen has succeeded in reducing digital transformation to its essence. Marc describes the digital breakthroughs in a clear and understandable way. This makes it a reference book, a guide and a source of inspiration at the same time for anyone who leads an organization or is responsible for digitalization or ICT."

– Erik Hietkamp, CIO / Director IT DAS Rechtsbijstand

"Nicely written book and interesting frameworks that are closely linked to the practical insights I gain every day."

– Martijn Ollefers,
Head of Digital Transformation for Allianz Benelux

"It is a beautiful book that is easy to read. It contains extremely useful advice and tools for anyone who is working on the digital transformation of their own organization or wants to start one. Digital transformation is a broad task that touches on people, process and technology - in that order. The book describes this strikingly, as well as the tasks and change areas that play a role in this and the importance of a well-thought-out transformation vision and strategy. The realization of that vision and strategy remains, precisely because of the often-radical nature of the change, ultimately the major task for every organization. In practice, digital transformation often turns out to be a bumpy road, with unexpected turns, necessary accelerations, beautiful views and, alongside the successes, also the necessary disappointments. Precisely at such times it is important to have a clear, well thought-out course. This book provides a number of very valuable ingredients."

– Arnoud van Petersen,
CIO Cyber Security Center

"Marc Beijen demonstrates great expertise and experience. In a clear and convenient way, the coherence between the latest insights in the areas of digital strategies, business models and processes, emerging IT trends and change management aspects is presented. The practical examples help executives and managers to make the translation to their organization."

– Albert de Bree, IT Director Rotterdam School of Management,
Erasmus University

"A beautiful compilation that is insightfully articulated. Written from practice, of which I was allowed to be a part."

– Peter van Wijk,
Change & Program Director Aegon, The Netherlands

"The coherence of complex digital changes is made manageable in this book for strategy formation and execution. Understandable and necessary for every manager."

– Rob Heerkens,
Member of the Board of Pension Fund PGB

"After reading this book you know what to do to digitally transform your organization. It's an easy read and the active writing style makes it possible to quickly see the steps you need to take in this complex subject."

– Gert van der Kooij,
Chief Digital Officer Hoogheemraadschap van Rijnland

"I find it very relatable; the book gives a good overview of the current issues with concrete suggestions for a successful digital transformation. You can get started with it in practice in no time."

– Henriette Setz, Director of Operations - Essent

"The book is a kind of Swiss Army knife in which relevant digital developments are sketched and then practical models, approaches and tips are provided on how to deal with this in your organization."

– Arie van Burgel, CEO of Univé South Netherlands

"Digital transformation can no longer be ignored. Both in the way we work and how we shape the world in these challenging times. But to realize this potential, a different view of your organization is needed. This book makes this complex task tangible, with concrete methods and solutions for taking steps. This makes it a good basis for any organization going through this transformation."

– Erik Oostwegel, CEO Royal HaskoningDHV

1 Digital revolution in the 21st century

1.1 CHALLENGES FACING ORGANIZATIONS

The world is changing rapidly. Organizations from all sectors are facing enormous challenges. Whether it's retail, healthcare, financial services, transportation or government, radical changes are occurring everywhere. The needs and behavior of consumers are fundamentally changing. They demand faster response times, transparency, excellent value for money and 24-hour service. Customers are increasingly organizing themselves, becoming more assertive and less loyal to suppliers and service providers. And customers want to be sure that doing business is safe.

New organizations are appearing on the market with new business models, positioning themselves ahead of the original service provider in customer contact. They steal your customer contact and push you into a supplier role with marginal revenues. Your position as the original service provider is lost forever! How will you respond to all this?

At the same time, technological advances are accelerating at a dizzying rate for organizations. This in turn raises new questions. What gives value to customers, employees, society and shareholders? What should you keep doing and what should you say goodbye to? What opportunities are you going to take advantage of in terms of market approach, business models, efficiencies and technology? What choices do you need to make?

The speed of change in society and technology means that organizations need to dramatically increase their flexibility, resilience and pace. An agile way of working certainly contributes to this but more is needed. The biggest obstacles for corporates, who have been in the market for a long time, are legacy problems. Not only of outdated systems and technology but also of associated processes, organization, practices, even culture. How do you ensure that this legacy does not block your strategy? How do you create overview and insight into which measures you need to take; which projects can continue or where you need to take drastic action?

The Greek philosopher Heraclitus proclaimed it long ago: *panta rhei*. This means, "everything is in motion," freely translated, "change is the only constant." This requires a trade-off between renovation and improvement of the existing, often side-by-side. So, renovate with the store open. How do you ensure that your organization smoothly transitions to the intended situation? In such a way that customers do not experience any negative effects and the organization retains control? And do you ever reach that intended situation, or is it an ongoing journey to Utopia?

1.2 ASSIGNMENT FOR MANAGERS AND EXECUTIVES

In today's society, digitization is an important factor to stay in business. More than ever, organizations will have to think and make choices about how digitization can create and contribute value to product offerings, services, chain collaboration and business operations. Optimal use of modern information technology can make the difference between decisive competitive advantage, survival and loss. In addition to focusing on the short term, one must also look to the future: what structural changes can you expect and how do you respond to them? Which (new) technology offers value to stakeholders? What opportunities do you seize in market approach, efficiency improvement and technology? Is what you want feasible? And how should you organize digital innovation?

Dependence on digital technology is increasing daily. If online banking is not available, the world quickly comes to a standstill. A failed ERP implementation can mean the end of your business. Developing digital channels too late can result in your customers switching to competitors. Investments in digital technology are, therefore, a significant part of the budget. Digital innovation is no longer a priority, it is a necessity. So, it requires structural attention from, and control by, managers. Every manager and director must therefore have the necessary knowledge, insight and decisiveness regarding digital technology. You can't leave this to specialists in your organization, or outsource it to a supplier or consultant.

Managers and executives need to know and understand the world of robotics, IoT, big data, blockchain, machine learning and artificial intelligence. Make the right choices from a sharp view of market developments, technological possibilities and business perspective. And having a clear picture of how to organize and manage this transition successfully. This is called digital transformation: an integral business transformation in which digital technology plays a dominant role. The goal is to create an agile organization and to use the opportunities, possibilities and expectations of digital technology in the utilization of data, the optimization of processes and with the focus on the customer. Digital transformation starts with the formulation of a resistant and successful *digital strategy*, a strategy in which digital technology is the main driver of change.

Literature on digital transformation has appeared with regularity in recent years. Often with promises and (often recurring) success stories about disruptive business models, for example in tourism (Airbnb, Booking.com), transport (Uber), media (Spotify, Netflix) and retail (Amazon, Alibaba). While we can indeed learn a lot from this, in practice it appears that organizations are struggling immensely with digital transformation. Research firms such as IDC predict that global spending on digital transformation is expected to reach around $2,300 billion by 2023. However, several studies show that only a minority of all initiatives will succeed. Why is it so much harder than described? What are the reasons for this?

There are answers to this. First, the success stories are often nicely polished; little or nothing is written about failures (and there will be many more). And that is a pity because it would provide many more and better insights and learning moments. Secondly, the case descriptions are almost always about newly established companies that have the "luxury" of a greenfield environment, which does not detract from their achievements. This contrasts with the established order, the existing corporates, that usually carry an opaque amount of ballast from the past. They are burdened with complex products, historically grown business processes, outdated information technology and even outdated competencies and culture. Complexity over which no one has the overview anymore. And we all know unlearning something is much more difficult than learning something new.

The digital vortex

1. Media & entertainment
2. Technology
3. Retail
4. Financial services
5. Telecoms
6. Consumer goods
7. Education
8. Business services
9. Tourism
10. Manufacturing industry
11. Transport & logistics
12. Real estate
13. Healthcare & pharma
14. Energy & utilities

The Global Center for Digital Business Transformation describes the "digital vortex"[1]. A vortex illustrates the disorder and complexity of digital disruption. Some industries are hit faster and harder by digital disruption than others. Sectors in the middle of the vortex are hit the most, while the effects on the outer edges are still light. But the vortex exerts a rotating force. As a result, sectors such as healthcare, manufacturing and energy are also drawn into the vortex. Sectors that are the first to be affected still have a great deal of work to do in terms of efficiency, convenience for the customer and transparency. More and more organizations are having to deal with the disruptive forces of digital disruption. Leaders of these organizations are increasingly aware of this, but a large proportion of them are still taking little or no action.

How will you approach digital transformation? Why, where and when are you going to work on it and how does it connect to the goals you have? How do you create an overview of the whole and determine what is relevant to you? How do you divide scarce means and resources between maintaining the current operations and implementing structural improvements and radical innovations? Do you organize renewal *within* the existing organization or deliberately outside it? But shouldn't you then integrate this at a later stage? Or do you, upon proven success, let the existing slowly die out? Which customer relations, services and business processes will be improved by all this and how does this relate to ongoing change processes? Where should (dis)investments be made? What choices must be made from the various change options?

1 "The Digital Vortex in 2017: It's not a question of when"; a four-part series of articles on digital disruption, by Professor Michael R. Wade.

These are just some of the practical issues that require a solution, it has become enormously complicated. Nobody has the overall view anymore: if you pull one string, others move with unknown consequences, everything is connected to everything. How do you get out of this? The core of the solution is making complexity manageable again. By applying structure, disentangling things and considering them holistically, in context. By distinguishing main issues from side issues, using the 80/20 rule and setting clear priorities. That is what this book is about.

1.3 DRIVERS FOR CHANGE

Making complexity manageable again starts with properly defining the problem. Or rather: the opportunities and challenges you are facing. This starts with recognizing drivers for change, the search for the WHY.

> **Reading tip:**
>
> *When digital becomes human*; Steven van Belleghem, 2014

Many drivers - in the sense of triggers, motives and incentives - can be identified that provide impetus for digital transformation. These have been clustered into six groups for ease of reference. However, these are not entirely isolated, there are interdependencies and they will interact:

1. **Emergence of new markets and business models**
   Organizations, machines and people are increasingly interconnected, forming a web of digital connections. These real-time interactions and data flows create new values that enable new products and services to be delivered more quickly and efficiently. There is a shift from traditional, technically closed infrastructures to open ecosystems that break down organizational barriers. This still sounds somewhat abstract but think of booking a flight with an airline: in a few clicks you have also purchased a hotel, rental car and tickets for the amusement park. Within a minute it's settled, and you have confirmations from all the organizations involved. These transactions not only involve the suppliers of what you purchase (airline, car rental company, amusement park) but also a digital platform, one or more insurers, a payment service and your bank. Organizations are changing the way they operate by becoming part of this "connected economy". Value chains are being unbundled, traditional business models are transforming to digital platforms at the top of the chain or to "low-cost" providers of partial solutions, more at the bottom of the value chain. The boundaries between industries are blurring, competition is becoming asymmetrical: companies are competitors and partners at the same time. Links in value chains are made redundant and pushed away or, on the contrary, new forms of intermediaries appear.

2. **Customer needs and expectations are fundamentally changing**
   The digitalization of society has led to an increasingly smaller world and this trend is continuing. It is becoming easier to find each other, regardless of distance. Organizing meetings between large groups of people has even become a matter of mere hours, if not minutes, due to social media. The masses can exert a great deal of influence, which can have both a positive and negative effect. Customers are becoming more vocal and demanding. The loyalty of customers to suppliers and service providers is decreasing, thresholds to switch are hardly there anymore.

3. **The changing relationship between man and machine**
   Digital developments have led to an information age in which new technological advances lead to new uses of people and machines. Developments such as "human enhancement", a collection of technologies that improve human productivity or abilities, facilitate our daily lives. This can focus on intellectual, physical or emotional abilities. The internet of things (IoT) is a network of physical objects such as cars, machines, household appliances and the like, which use sensors and software interfaces (APIs) to connect to, and exchange data over, the internet. Machines are becoming more intelligent and can take over human tasks more and more easily and quickly. As a side effect, many professions will fundamentally change or even disappear.

4. **New technological developments offer new opportunities**
   Reality and the (virtual) digital world are increasingly merging, causing people to see and experience society differently. Technology is increasingly diffused in the physical world, boundaries between digital and physical are blurring. Applications of technologies such as "augmented reality" and "voice recognition" are beginning to take an increasingly prominent role in society. Blockchain technology can improve efficiency and transparency in the value chain, enable secure mutations and minimize the role of third parties. Artificial intelligence (AI) combined with automation and robotics offers opportunities to dramatically optimize processes. Digital twin technology, a digital replica of a living or non-living physical entity, can be used in product development, calculating what-if scenarios, production improvement and maintenance. These are just a few examples of the enormous potential that digital technology has. And this development is continuing all the time.

5. **Growing regulatory, privacy and ethical requirements**
   A driving force of an entirely different order but certainly no less important. Organizations must increasingly weigh up the pros and cons of quickly keeping up with new technological developments on the one hand while considering the opinions (or preconditions) of the outside world on the other. This gives organizations less and less freedom compared to the past. These changing

relationships between society and organizations mean that an organization must make important trade-offs. Is everything that is technologically possible always desirable? The reality is that any new technology can be used for good or bad. For example, gene technology, privacy violations and cybercrime, so how do we deal with this? Robotization can eliminate entire professions, but do we want that? This calls for new principles and rules for organizations. A clear balance must be found between responsible business and making use of the latest digital opportunities, which are not always without risk. At the same time, we see that national and international legislation often lags digital developments. In many situations, regulations are not redesigned but instead (to the despair of entrepreneurs) grow exponentially. All this affects how we price and sell products, how we manage risks and how we organize our business. Again, this increases complexity, reduces returns and requires strong adaptability.

6. The increasing value of data

Data is typically called the most important asset of today's economy. As a result of ongoing digitalization, organizations are generating ever larger amounts of data. It is a major challenge to obtain the appropriate insights from the masses of data so that new products and services can be developed. The ability of organizations to use these large amounts of data intelligently will cause a change in the relationship between organizations and society. Data is also a driver for further digitalization such as AI and machine learning. Advanced customer analytics such as behavioral and predictive analytics are increasingly becoming a necessary core competency. For example, applying statistical algorithms and machine learning techniques enriches insurers' propositions and reduces or prevents risks. Large amounts of data (e.g., generated by IoT and wearables) can be analyzed to identify customer patterns and form the input for personalized insurance products and risk management.

Are we complete with these drivers? No. Who would have predicted in 2019 that the coronavirus pandemic would break out with such a huge impact? We can now see that this will make organizations focus even more on creating digital relationships with customers and partners, and further digitizing work environments for employees. You cannot predict the future, but you can prepare for it, at least better than your competitors. You can have an influence on what the future of your organization will look like. Organizations that master this are the winners of tomorrow.

2 Digital transformation in perspective

■ 2.1 PHASE MODEL FOR DIGITAL TRANSFORMATION

> **Reading tip:**
>
> *The Alchemy of Growth*; Mehrdad Baghai a.o., 2000

No two organizations are the same. Even if organizations are remotely similar, are "pure" competitors, sister companies or business units within a larger concern: there are always differences in market approach, products, business processes, culture, management style and deployed technologies. The starting point is different for everyone. Therefore, there is no generally applicable blueprint for successful digital transformation. It will differ for each situation and needs to be tailored to the state of the organization, and the ambitions and goals of the management.

A tool for this is the phase model for digital transformation, inspired by the "Three horizons of growth framework" of McKinsey[2]. In this - highly simplified - model of reality, three phases are recognized in which an organization can find itself. We must realize that digital transformation is not a one-time thing. It is not something that can simply be implemented by purchasing the latest technology. It is a continuous process that is constantly evolving, with no end in sight. More phases will undoubtedly follow in the future.

We will mainly recognize organizations that are active in **phase 1** from our past. But you will be amazed at how many are still in this phase, or at least exhibit a good number of characteristics from this phase. They have a strong product focus and think from the inside out. The customer does not always feel treated as a customer. Products are often complex and not modular. Sales and service take place through a limited number of channels that are not, or hardly, attuned to each other. The organization is hierarchical, task-oriented and often functionally structured. There

—

2 "Enduring Ideas: The three horizons of growth", McKinsey Quarterly, December 1, 2009.

is a formal customer-supplier relationship with external parties. The information provision is focused on transaction processing, whilst in terms of technology the systems can be characterized as administrative, monolithic silos that are set up as a single entity per product group or business function. With a lot of customization and dependencies on internal and/or external specialists. The IT organization is run like a factory, "you ask, we deliver".

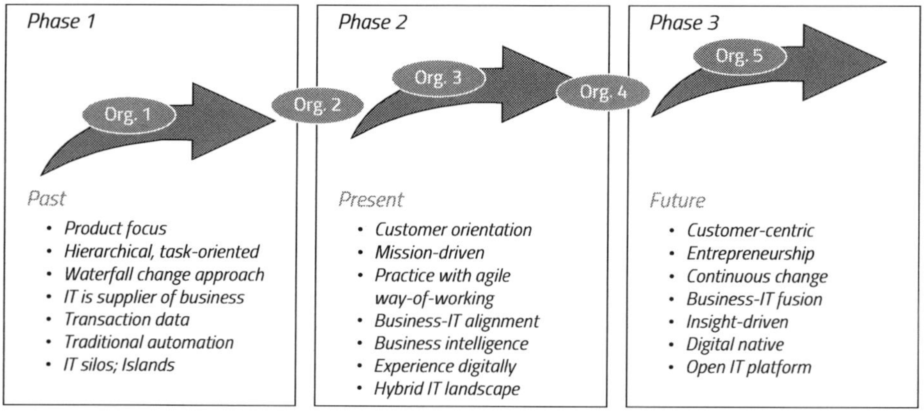

Figure 1 Phase model for digital transformation

In **phase 2**, there is customer orientation. Besides offering products, explicit attention is paid to service (in the manufacturing industry this is called "servitization"), with customer satisfaction being an important indicator. Products are simplified and modular. A CRM layer is placed over the various product groups and an integral customer view is built up. Products and services are offered through multiple channels (cross-channel), with self-service via the internet playing an important role. The organization is mission-driven, non-core activities are outsourced. Reporting and data analysis take place using business intelligence software. The IT landscape is hybrid: a mix of older and newer systems, decoupled and integrated again via middleware. There is business and IT alignment: demand for, and supply of, IT are well matched. Agile working is applied in various places. New roles such as product owner and scrum master are being introduced.

Phase 3 is how we now view the new world. There are already organizations in our midst that meet these characteristics. Everything is about creating relevance and an excellent customer experience. Propositions are fully modular and personalized to customer needs. Distribution takes place seamlessly across channels (omnichannel), the customer can do business anywhere and anytime. The organization acts in a network, collaborates with business partners and entrepreneurship is present everywhere in the organization. There is no longer a distinction between run and change. Multidisciplinary, autonomous teams work continuously on processing feedback from the market, renewing value propositions and optimizing operations.

IT is part of the business and acts as an accelerator in realizing business ambitions. Data is one of the most important assets and, supplemented with smart algorithms, is used to optimize customer journeys. Decisions are made based on facts and predictive analyses. They have an open IT platform: a mix of in-house and external systems, which are decoupled by means of standardized interfaces, APIs. Open-source software, open languages, frameworks and libraries are used, as well as cloud solutions that can be scaled up easily.

Of course, when you read it that way everyone wants to get to phase 3 quickly. That's where most of the recently published books and publications about digital transformation focus upon. But it's not that easy. An organization acting in phase 1 cannot simply skip phase 2. But it can, with the right strategy, accelerate that step, gain experience and prepare well for phase 3. Some organizations still make a lot of money in phase 1 or phase 2, especially internationally operating companies in countries where the digital revolution is still in its infancy. So, should you throw that overboard? And is the industry in which your organization operates ready for phase 3? But look again at the digital vortex in Section 1.2: sooner or later the sector in which your organization operates will also be drawn into the vortex.

In any case it is important to make a good (and honest) estimate of where your organization is in the model (that can also be "in-between") and what a feasible and manageable ambition is for the coming period. For larger organizations, it is better not to do this for the organization as a whole, but for each component. Because that is what we started with in this paragraph: no organization is the same, not even business units within a larger enterprise.

2.2 NOT ONE BUT TWO DIGITAL TRANSFORMATIONS?

From the phase model we can see that there are two different digital movements. The movement from phase 1 to phase 2 is aimed at turning a product-oriented organization into a customer-oriented one, including the drastic simplification of everything in your organization: products, processes, working methods, management and IT landscape. The movement from phase 2 to phase 3 is aimed at realizing the company of the future: renewing, disrupting, transforming.

Many managers are (fortunately) extremely ambitious and want their organization to be ready for phase 3 as soon as possible. But practice is obstinate, often the organization is still struggling with the transition from phase 1 to 2.

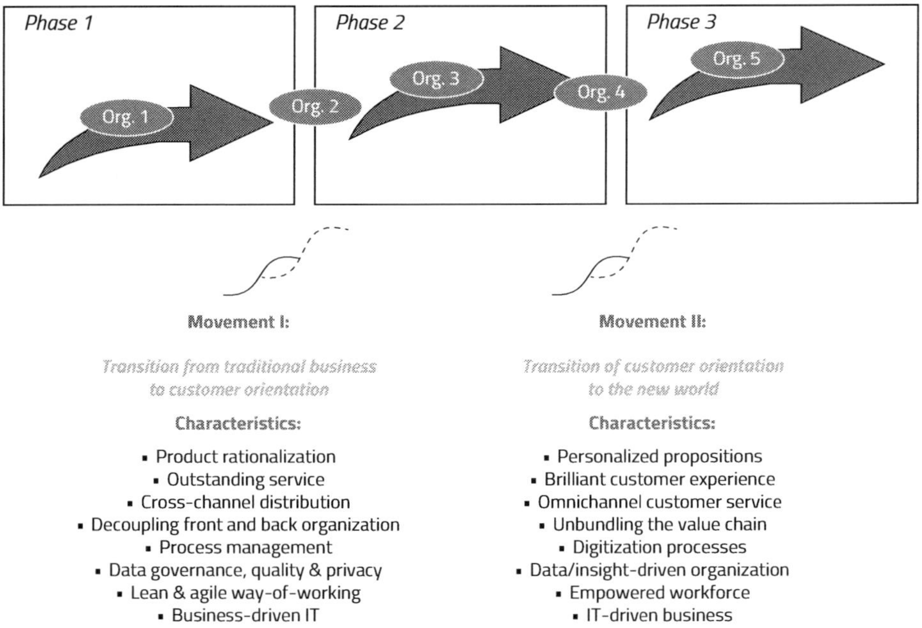

Figure 2 Two digital transformations

In order to move on to the next stage, it is important that there is some balance of maturity with respect to the stage characteristics. For example, you cannot make the move to phase 3 if you are not in control of data. If the quality of data is not in order, if it is unclear who has what responsibility for data or privacy is insufficiently regulated, then you cannot trust data in predictive analyses and reports. What's more, the customer will soon abandon the belief in your good intentions.

You cannot go to the new world if the culture and management style of the organization are not ready for it. For example, if managers do not delegate but want to do everything themselves, they have insufficient confidence in their people and steering is not done on the WHAT but on the HOW.

So, it is better not to focus on one or a limited number of phase features. Everything must be in order before you to make the step to the next phase. But you can't do everything at once, so you will have to make some sharp choices about what, how and when to tackle things.

Reduction of complexity

You can forget about phase 3 anyway if you fail to remove complexity from the organization. Simplify as much as possible the products, processes and IT in your company. Then you create a more flexible, agile organization that can respond faster to changing customer expectations. Stop pretending that your company is so unique; the reality is that most of it is pretty standard. Where you are truly distinctive, which will be at most 10 or 20%, you may well come up with customized solutions.

A new company can start right away in phase 3, whereas the established ones don't have that luxury. But evolving from stage 1 to stage 3 can take too much time that you don't have and leave too much room for the excuse "yes, but we are not yet able to ...". So there needs to be pressure, you can't afford to take many years. You must accelerate phase 2 and make "no-regrets" choices that will promote go-around to phase 3. In everything you do, when designing the situation to be achieved, you must integrally design the required digital capabilities that belong to phase 3. And start experimenting, gaining experience and learning. This will ensure that you can proceed to the new world at an accelerated pace.

The following overview shows the main results of the two movements described. These are useful as a checklist; delete or add what is important for your organization. If you have sufficiently achieved the results relevant to your organization in the first movement, then you are ready to steam ahead into the new world.

 **Movement I:** *Transition from traditional business to customer orientation*	 **Movement II:** *Transition of customer orientation to the new world*
Modular products • Products and services have been rationalized, simplified and modularized.	**Personalized propositions** • Market propositions consist of bundled partial solutions and are put together in a personalized way based on customer needs.
Excellent service • Customers experience structurally better service in presales, sales and aftersales, the customer can do business through multiple channels, there is always and everywhere the possibility of self-service, there are hardly any "wasted calls" and customer requests are solved "first time right".	**Brilliant customer experience** • You have a distinctive position in the market in terms of customer service, a seamless customer experience across all channels. Customer service is focused on customer events and customer journeys, service is maximally digital (chatbots, web, app, ...) with a human touch: where it is needed human interaction makes the real difference.
Efficient processes • The front-end of your organization (distribution and customer service) is decoupled from the back-end (production and administration). • The front-end is focused on distinctiveness, added value and exclusivity in service provision. In order to be able to deliver this, flexibility in the organization is required, so that specific target groups can be quickly accommodated. • The back-end is focused on creating scale, cost reduction, zero defects and risk management. Processes are standardized and automated where possible. Input controls are as early as possible in the process chain. Defects are organized into separate processes. • There is clear process responsibility, control of customer processes and integrated case handling. • Processes are governed with KPIs, the organization learns from customer reactions.	**Smart, digital processes** • The different roles in the value chain have been recognized and unbundled. Choices have been made about which role(s) your organization takes on itself, or for which roles it works with partners. • Business processes have been fundamentally redesigned. Not with the cheese slicer but truly greenfield. There are no compromises, sub-optimization, exceptions and unnecessary complexity. • Where it adds value, processes have been digitized and supported by smart technology: chatbots, RPA, machine learning, AI, blockchain etc. • People are no longer deployed on simple, repetitive processes but on work with real added value for customer and business.

 **Movement I:** *Transition from traditional business to customer orientation*	 **Movement II:** *Transition of customer orientation to the new world*
Data in order • Data quality and data privacy are in order; customers can trust that their personal data is protected, and integrity and confidentiality are ensured. • Relationships are uniquely identified. The authentic source and owner of all data are known. • Data is leveraged for a clear and comprehensive customer view to support sales and service. • Data is used for reporting and (market) analysis.	**Data-driven organization** • The organization is data-driven and is fully trusted by all stakeholders. • Data generates a continuous stream of insights and is the input for predictions, scenario analysis, automated decisions and digital innovations. • The organization is able to use (the value of) data for customers, chain partners, employees and society. • Data is used in an open environment by all market participants.
Competent employees and managers • Employees are masters of their craft. They have a strong sense of responsibility, are team players, proactive, curious and eager to learn. Digital thinking has been adopted. They have a strong IQ and EQ. • Managers are visionary, authentic, communicative, and think and act from the outside in. They can distinguish between main and side issues and delegate clearly to employees.	**Internal entrepreneurship** • Employees are creative, act on facts, are decisive and enterprising. They learn daily from both successes and failures. They are "digital natives" and have an enormous passion for the customer. • Managers give space and trust to employees, are facilitators, connectors and lead by example. There is room for experimentation. Managers give direction (in advance) and clarity on strategy, policy and frameworks.
Grip on change • Changes are derived from the business strategy and are managed through change portfolio management. • Business cases are prepared and monitored; if not successful enough, projects and change initiatives are discontinued. • Change roles and consultation structures are in place; there is a grip on change.	**Agile and resilient organization** • The organization is focused on maximum flexibility (adaptability) and innovativeness; change is in the blood. • Multidisciplinary value teams have been established that, from idea to realization, work continuously on improvement and innovation. • We work according to the principle of learning by doing. The next step is taken, based on insights gained from practice.

 **Movement I:** *Transition from traditional business to customer orientation*	 **Movement II:** *Transition of customer orientation to the new world*
Effective IT landscape • Information systems are composed of several decoupled layers: process, data, functional and interaction layers. • There are no redundant IT solutions for the same information needs: one-time configuration of systems with the same functionality. • Standard software packages are used for standard processes and this applies to a large part of the organization. Processes are adapted to these solutions so that developments (releases) from the supplier can be followed. • CRM functionality is available to anyone with customer contact, anywhere, anytime. • There is integrated process control using BPM (business process management) and WFM (workflow management) solutions.	**Open IT platform** • The IT landscape is based on an open IT platform: a mix of in-house and external systems. • SaaS solutions (software as a service, an application offered as an online service) and PaaS solutions (a platform offered as a service that allows applications to be developed, run and managed, without the complexity of building and maintaining the infrastructure yourself) are used. • The IT architecture is flexible, scalable and can grow with dynamic and changing environments. • Components in the IT architecture are decoupled through standardized interfaces (APIs - Application Programming Interfaces). • Information and functionality are easily accessible to all authorized users (employees, customers and partners). • Applications are designed and developed with the user and customer in mind ("digital native"). • Digital connections have been made to externally available information sources. • Maximum use is made of the benefits of the cloud.

2.3 FIVE DIGITAL BREAKTHROUGHS

Entrepreneurship in the 21st century requires rapid development, rollout and scaling of successful business models in different markets. At the same time, we need to nurture what has been built and successful in the past. You must simultaneously play chess with different strategies on multiple boards: maintain and optimize the existing business, achieve growth by creating more relevance for customers and simultaneously develop new business models, greenfield or through acquisition. You will need to build an agile organization, create synergy, experiment and reuse innovations.

Looking at the phase model for digital transformation, there is a common thread in the digital breakthroughs that need to be realized to carry through the ambitions and business strategy. These digital breakthroughs can be summarized in five themes:

data-driven organization, smart digital processes, brilliant customer experience, agile and resilient organization and new digital business models. All themes can be linked to your business strategy in one way or another.

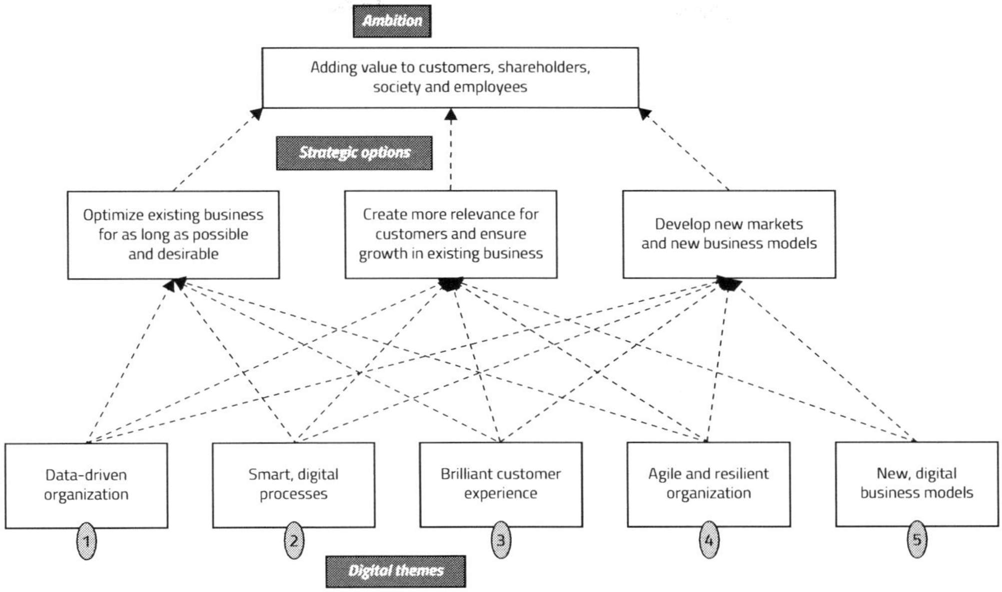

Figure 3 Digital breakthroughs in relation to business strategy

Depending on the stage your organization is in and the digital ambition you want to realize, the goals will vary by theme.

Theme	Goals movement I	Goals movement II
Data-driven organization	Available data of good quality, privacy is in order; reporting and simple data analysis	Insight-driven organization; data is used in the chain, value creation with data
Smart, digital processes	Controlled, uniform and efficient business processes	Processes are digital and supported by smart technology, people make the difference for the customer
Brilliant customer experience	Excellent customer service in sales, pre-sales and after sales	Distinctive, omnichannel customer experience, customers are involved in your organization
Agile and resilient organization	Capability for change in order; grip on everything that runs	Continuous change organization, focus on value creation, fast and agile, short-cycle, change is everywhere
New digital business models	Simplified and modular products and services, extend current business models	Market introduction and upscaling of digital platforms with potential, you are part of the 'connected economy'

Figure 4 Five digital breakthroughs linked to the movements

On every digital calendar, these topics will appear to a greater or lesser extent; they are relevant to almost every organization. But it is certainly not only these topics; depending on the situation, there will be more and other topics that are high on the priority list, for example:

- Effective distribution: further broadening, deepening and improving the effectiveness of your distribution channels.
- Control and compliance: ensuring that you are fully in control of all aspects of business operations, and that you comply with relevant laws and regulations ("license to operate").
- Digital DNA: developing knowledge, skills and culture, such that "digital" is in the capillaries of the organization, among all employees and all managers.
- IT security: protecting yourself as an organization from unwanted intruders, making sure your internal governance is in order, no one can "just" access everything.
- Sourcing: "make, buy or ally", determining what you will continue to do yourself (make), what is best to outsource (buy) and where you will work with partners (ally)? And if you outsource, is the cloud a good idea?

It will be down to you to discover which themes are relevant, so how should you do this? The next chapter provides the necessary tools. You would do well to add them to your digital strategy.

3 Digital strategy as a starting point

3.1 ELEMENTS OF A DIGITAL STRATEGY

As already mentioned, moving to the new world requires a sophisticated strategy. Digital transformation starts with formulating a solid digital strategy, one where digital technology ("push") interacting with creative, new business concepts ("pull") are the key drivers of change. A digital strategy is not something separate, it is part of the business strategy built around the digital capabilities of the future. The digital strategy defines the way in which digitization and the use of data improve products and services, customer service and business processes. It guides the development and adoption of new technologies and even new business models. The strategy is achievable and makeable: choices in the digital roadmap and the matching enterprise architecture are taken in such a way that they can also be realized.

The process of coming up with a digital strategy is interactive, iterative and creative. In addition to performing all kinds of analysis and thinking, it is important to include ideas from within the organization. After all, the best ideas are not only generated in the boardroom. Many will come from top people from your organization, who are close to the customer and know everything about processes that run through your organization. By structuring the results of the inventories and analyses, an overview of digital themes that are geared to your business is created. The five themes we mentioned previously, the digital breakthroughs, will undoubtedly be part of this. Next, prioritization and the actual selection process takes place. After considering the various scenarios, choices are made that lead to a digital roadmap.

The following elements are recognized when developing a digital strategy:

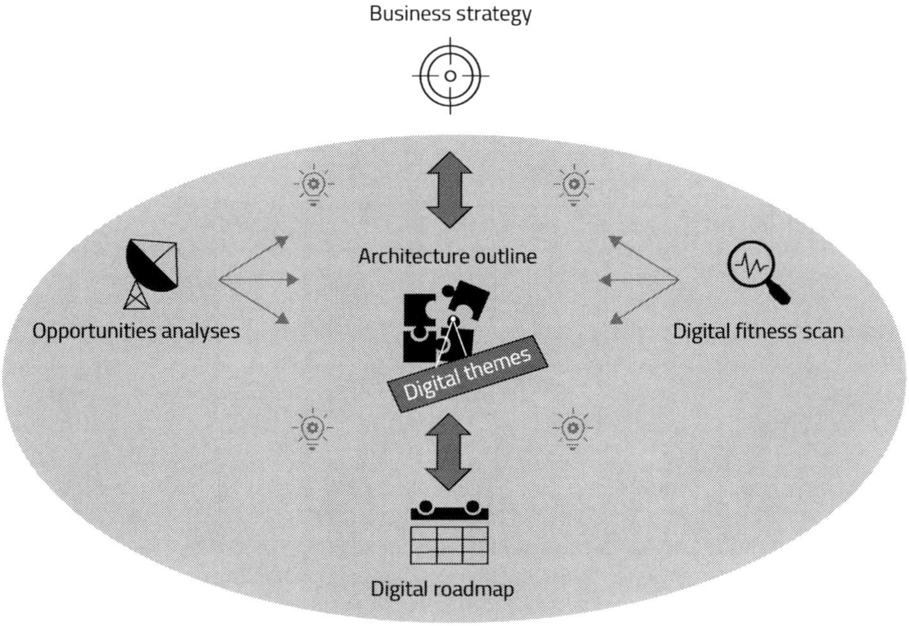

Figure 5 Elements of a digital strategy

1. **Business strategy**

   The business strategy gives direction to all decisions an organization makes both now and in the future. It formulates the dot on the horizon, your strategic vision, ambition and the goals of the organization to be realized. To achieve a good digital strategy, you need to have at least the following clear:

> **Reading tip:**
>
> *Start with why*; Simon Sinek, 2011

- What mission has your organization set for itself, what is its "purpose" and identity, what legitimizes its existence, what do you stand for? A mission is timeless but can be applied at any time.
- What is the vision, the inspiration of your organization? What do you want to go for as an organization, what does the desired dream situation look like?
- What (long-term) goals have been formulated in this regard? What are the tangible results you are aiming for in order to achieve your organization's mission and vision? To achieve goals, you need to make them SMART: Specific, Measurable, Acceptable, Realistic and Time-bound.
- What positioning does your organization choose in the market? What internal and external developments affect the mission, objectives and positioning?

The business strategy is, in turn, influenced by the digital strategy; it is a cyclical process. Therefore, the digital strategy is not something separate; it is part of the business strategy.

2. **Opportunities analysis**
   This concerns an outside-in analysis of the drivers of change introduced in Section 1.3. You look at what is going on in the environment and in society as a whole, and from that knowledge you deduce what you can and/or must do with it. You don't only look at current market segments and positions of the organization but also at adjacent segments, potentially new customer groups and alternative roles in the value chain. This approach forces you to think outside the traditional box. Key questions that need to be answered are:

   - What does the market demand from us?
   - What is our competitive position?
   - What developments are coming our way and what do we need to do about them?
   - How is the demand for our products and services developing?
   - Which positions in the chain provide the most value?
   - Which networks and partnerships are interesting?
   - What are the possible surprises?
   - Which product-market combinations will we optimize, and which do we see as growth businesses?
   - Which product-market combinations will we say goodbye to?
   - What new business models do we see for the future?

And for all these questions there is the overriding consideration of what possibilities and opportunities does digital technology offer?

3. **Digital fitness scan**
   The digital fitness scan is an inside-out analysis, an approach where the situation within your organization determines what the organization does in the market. You look at what the organization can do and from that knowledge at what capabilities you have; the strategic fit is determined based on the quality and competencies of the organization. The digital fitness scan is an assessment of the current versus the required digital capabilities (see the checklist from Section 2.2) and provides answers to the questions: what about the need for change, the willingness and ability to change, where are you now, what phase is your organization in, what are the digital core competencies and what needs to improve.

4. **Architecture outline**
   When developing a digital strategy, it is useful to draw up "architectural sketches": the translation of the business strategy into principles, guidelines and models about the organization of the business and the related digital technology. These architectural sketches map out both the existing ("as-is") and the desired

("to-be") situation and form a frame of reference on which to identify and hang digital themes.

The importance of enterprise architecture
We know architecture as an approach for designing a building but something similar applies to designing a (digital) organization. There are several types of architectures: process architecture, information architecture, application architecture and technology architecture. And often forgotten: the architecture of products, services and distribution channels. All these architectures come together in the enterprise architecture (EA).

The essence of architecture is to define and operate a set of generic design principles and decisions, supported by formal agreements (conventions, protocols, standards), from a particular vision. However, the reality is that an organization and its IT landscape are not, or are only partially, designed as a whole (in one coherent architecture); there is more of an organic growth. In the past, the starting points for choices in market propositions, design of business processes and development of information systems were often the circumstances in a limited area of the organization (a department, a product group, a region) at a certain moment in time, using the possibilities and technology that were available at that time. The consequences of the absence of an overarching architecture manifest themselves in unacceptable costs for management and maintenance, and excessively long development times for the introduction of new propositions or changes in processes and organization.

A well-thought-out enterprise architecture is necessary to realize a renewed business design that does justice to the new reality and can flexibly adapt to future developments. There must be an optimal connection between market propositions, business processes and supporting IT facilities. Organizations that are mature in information management and alignment between business and IT usually have their enterprise architecture in order. They connect propositions, processes that produce propositions, information that belongs to these processes, the application landscape and the technology deployed for it. If the strategy is adjusted, the architecture - and its implementation in business operations and information provision - can respond adequately and dynamically. Competitors are then quickly overtaken left and right and are left behind.

The purpose of an architecture at the level of digital strategy is not to provide a static and detailed construction drawing but rather has the function of an "artist's impression". So, no "big design"; after all, the world is only marginally designable. Such an architecture provides a vision and principles on one or more future scenarios, has a longer-term validity and is inherently subject to change and discussion. In this context, architectures are particularly important for communication with, and decision-making by, stakeholders in the organization and not directly intended as detailed implementation models in a technical sense, although they certainly provide direction for this.

5. **Digital roadmap**
   The digital roadmap is the guideline, with choices you select to go from today to tomorrow. Not extensive and linear but outlined, short-cycle and incremental; after all, tomorrow the world could look different again. The digital roadmap includes the changes needed to realize the digital ambition. Various changes can be tackled and executed via the traditional waterfall approach, many in an agile way or in a hybrid model. For each change, it will be necessary to consider which method of working suits it best. An important point here is: it is never an IT change, it is always a business change. Digitalization affects the organization deeply and has structural consequences for the work of employees and the relationship with suppliers and customers. This approach ensures that the right knowledge and skills are present, and that acceptance and implementation will be easier. In this way, the possibility of success is greater than with the "narrow" approach taken when considering change from purely a technological perspective.

3.2 FRAMEWORK FOR BUSINESS TRANSFORMATION

The Business Transformation Framework[3], introduced in this section, helps to formulate a digital strategy and to identify all the changes you need to make in order to implement a digital transformation. The framework provides insight into the complexity, helps to structure, shows coherence, and provides tools for deciding what to take on. In other words, it is an indispensable approach to get and keep a grip on digital transformation.

If we look at organizations from a distance, we see that they consist of roughly the following four pillars or dimensions:

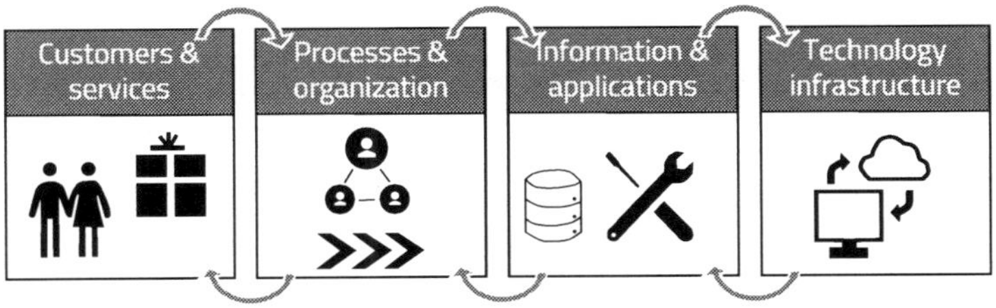

Figure 6 The four pillars (dimensions) of an organization

3 First published in 2001 in the book *Strategic deployment of ICT; a guide for business information management*, Marc Beijen, Eric Broos and Etienne Lucas.

1) **Customers and services**: The starting point is the customer need, often linked to events in the customer's life (live events such as studying, getting married, moving, retiring, etc.). Customer events and customer needs are the triggers on which we design and implement our products, services and communications. We deliver value propositions proactively and at the customer's request, which are personalized and relevant based on their profile and needs. Value propositions must be viewed from the perspective of the requirements and desires of the environment. This concerns the total properties of a product or service to be delivered in relation to the sales market: function, pricing, distribution, quality, service, image, etc. (the so-called "marketing mix").

Marketing mix

Philip Kotler distinguishes the so-called marketing Cs (formerly the classic marketing Ps)[4]:

- Convenience (instead of "place"): this is about convenience, atmosphere and experience around products and services.
- Cost to consumer (instead of "price"); the actual total cost that the customer must commit to in order to obtain a product/service.
- Communication (instead of "promotion"); the interaction (two-way traffic) with the customer, at several moments, preferably in a long-term relationship and not only at the contact moment.
- Customer solution (rather than "product"); thinking from the perspective of what the customer wants rather than unilaterally determining a product/service.

Note that this is equally true for government and not-for-profit organizations. They too have an environment that places demand on their products and services, which require thinking about in terms of how to meet them.

2) **Processes and organization:** Value propositions are produced by processes involving multiple parties. Processes are chains of activities that are necessary to develop, produce, offer, distribute and provide aftercare for value propositions. The process view provides insight into which (internal and external) parties add value at which moments in the realization of the final product. Processes are preferably executed digitally, directed and completed by competent employees who have the right skills, attitude and behavior to solve problems when it really matters. Processes are organized into an organizational structure: business units, departments and/or teams.

4 Phil. T. Kotler, *Principles of Marketing,17th edition*, July 2017.

3) **Information and applications**: Processes and people in the organization, as well as customers and business partners, are supported by information and applications such as an integrated customer view, contact registration, workflow management and management information. In every organization many dozens of information functions can be identified. They are the building blocks of the information provision and provide communication with the environment and the management of the organization. In addition, applications of information technology can be identified in the means of production (for example, advanced technology in trains), in products and services themselves (for example, in smart thermostats) or as independent marketable products (for example, navigation solutions).

Example of digital technology in trains
Trains are moving computers, exchanging vast amounts of data in real time with the systems of our national public transport operator and other parties: this includes reliable and accurate information about train location, configuration, speed, seat occupancy and equipment condition, among other things. This data is used to inform passengers about departure and arrival times and seat availability, to support advice to train drivers and to provide technicians with maintenance information.

Digital twin platforms, using complex event processing and edge computing, are used for new train types and have been extended to include new services for customers and internal users. This concept lowers the TCO (Total Cost of Ownership) compared to train manufacturers' "silo solutions" and simplifies the commissioning of new trains.

4) **Technology infrastructure**: This includes all technical facilities required to develop, operate and manage information systems (databases, networks, standard software, middleware, hardware). This technology can be managed in-house (on-premises) or cloud solutions can be used, effectively as a form of outsourcing.

When the above four pillars of an organization are plotted against the concepts introduced earlier - business strategy, architectures and roadmap - the framework for business transformation emerges.

There is consistency between all parts of the framework, from top to bottom and left to right and vice versa.

From top to bottom (and vice versa): The business strategy formulates the ambitions and goals. Starting points are guiding principles, connected to this strategy. In the layer "architectures", the principles are translated into (scenarios of) the future situation in the four columns. By comparing this with the existing situation (fit/gap analysis), we get a clear idea of the actions that must be carried out to move

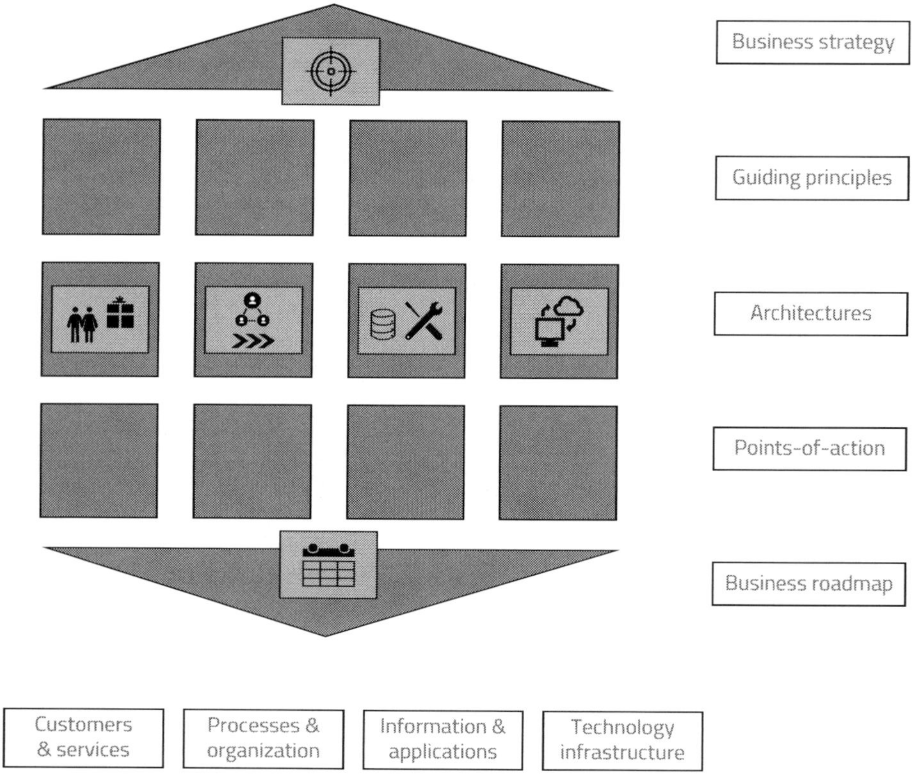

Figure 7 Framework for business transformation

towards the desired situation. These are bundled into work packages, projects or "epics", each of which realizes part of the strategy and thus leads to business value. This is the top-down approach. But the other way around, bottom-up, is just as important: the best innovations don't always start in the boardroom. Within your organization, fantastic ideas developed by employees undoubtedly exist, whilst new initiatives are already being implemented, each of which can have an impact on the business strategy. The two approaches interact with each other.

From left to right (and vice versa): Customers and services are the starting point for our actions. Processes and organization are aligned with them and designed accordingly. Information and applications are needed to execute processes and support customers, business partners and employees. The technology infrastructure supports all of this. Theoretically, there is a logical sequence between the four dimensions. In practice, there is always a strong interaction between the various topics. Technology from the fourth column offers new possibilities and opportunities in the left columns, which would never have occurred to anyone thinking merely "from left to right".

The framework helps to translate the business strategy into a digital roadmap ("vertical alignment"). In addition, it ensures that all dimensions are aligned: customer needs, value propositions, business processes, governance, people, competencies, information, functionality and technology ("horizontal alignment"). Using the framework, you can ensure that all parts of the "organizational system" are aligned. Success is only partly achieved with separate parts; it is mainly about the cooperation and "fit" between these parts. The coherence is crucial.

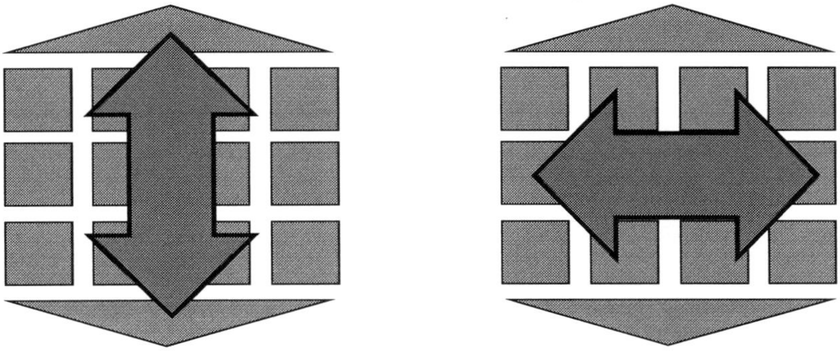

Figure 8 Vertical and horizontal alignment

The framework helps in several ways:

Reading tip:

Business Transformation Framework; Sjoerd Staffhorst, Remco Bekker a.o., 2016

- It provides overview, structure, clarity and transparency to all stakeholders about the desired situation and how to get there. Understanding complexity makes complexity more manageable.
- The framework enables digital themes to be clearly defined, positioned (among themselves) and prioritized. And it then forms the frame of reference for the review of digital initiatives on goals, results, scope, dependencies and risks.
- It ensures that all investments in digital transformation are focused on adding value for customers, employees, shareholders and the environment, and simplifies decision-making about resources (money, knowledge and capacity).
- The framework provides insight into feasibility and manufacturability ("what do we want" versus "what can we do").

The business transformation framework provides a guideline for deriving and shaping the digital strategy. The results of all elements of the digital strategy as discussed in Section 3.1 can be included in the framework.

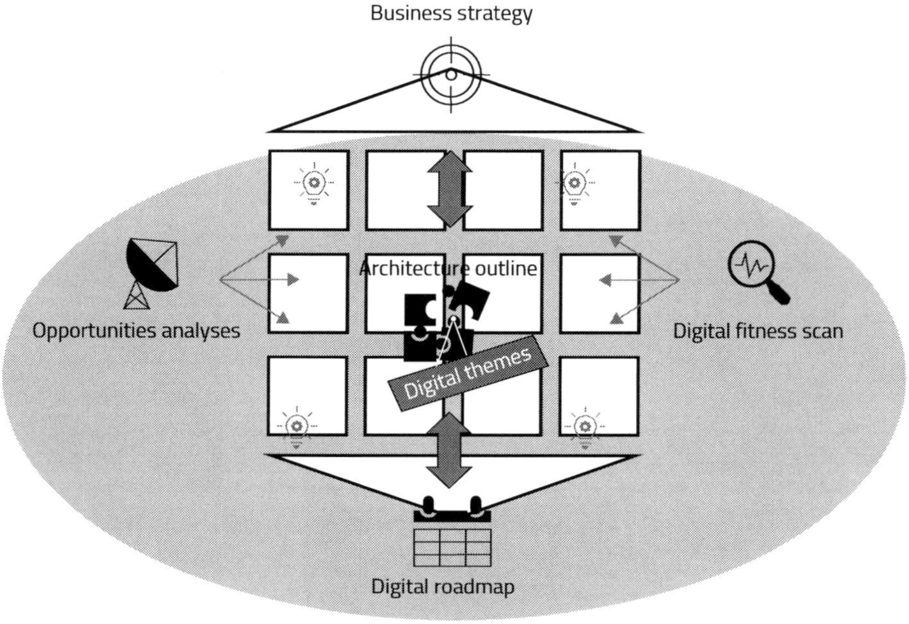

Figure 9 Business transformation framework as a guideline for digital strategy

3.3 DIGITAL BREAKTHROUGHS PROJECTED ONTO THE FRAMEWORK

We can also project the five themes, the digital breakthroughs, onto the business transformation framework. Each breakthrough can be positioned dominantly in one of the columns or rows of the framework, but it also sets requirements or creates opportunities in other columns or rows. By going through the framework for each theme you will get a clear idea of what needs to happen in your organization to realize the goals and ambition in that theme.

For example: data-driven work is a theme that is primarily positioned in the third column (information and applications). Technology in the fourth column is needed to be able to work in a data-driven way, for example a data lake with import and export functions. But it also makes demands on the second column. You need data governance such as policies on data quality and privacy, agreements on data ownership and clear decision-making processes. You also need competent employees (data analysts, data stewards). You can use the data to make better predictions, do more effective marketing or apply dynamic pricing.

Something similar is true for all themes. Moreover, there are dependencies *between* the themes. Data-driven organization will contribute to improving the customer experience, for example by being able to offer personalized service. And it will promote smart, digital processes, for example as input for AI and machine learning. Another example: to successfully develop new, digital business models you need an agile organization, brilliant customer experience, digital processes and data-driven organization. Again, this shows that coherence is crucial.

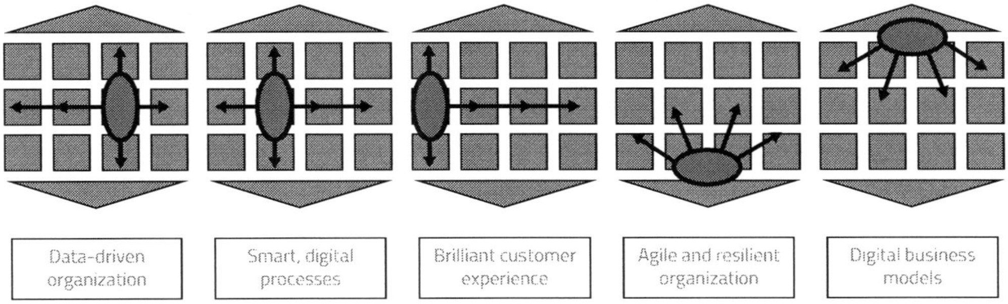

Figure 10 Digital breakthroughs projected on the framework

Not only the five themes, but also all other necessary change activities in your company can be traced and depicted on the framework. In this way a total overview of your organization's change assignment is created.

What about the fourth column?

You will notice that no digital breakthrough is defined in the fourth column, technology infrastructure. Why is this the case? We're talking about smart use of digital technology, right? It is precisely the capabilities of technology in the fourth column that enable digital transformation. So, shouldn't a digital theme be formulated on that?

The reason is that we are primarily talking about *the application* of digital technology and that is in the other columns. Technology must support this and offers new possibilities and opportunities. Without application it is of no use to you. Technology infrastructure is also less and less of an issue. Solutions are composed of standard components; programming is less and less necessary. The same applies to hardware, after all you are using the cloud. Infrastructure comes like water from the tap, as required. You only need to think about its place in the IT landscape and arrange connectivity, security and management.

Nonetheless, there is also a substantial change task to be recognized in the fourth column. The IT infrastructure will have to migrate to an open IT platform, consisting of decoupled components that can "talk" to each other and the outside world through standardized interfaces (APIs). There will be a need to migrate to the cloud, where this adds value, instead of buying and managing everything yourself (on-premises). And the IT supply chain, from planning to management, will want to be automated to a large extent. The Open Group, in which consumer organizations, IT suppliers and academic institutions participate, has developed IT4IT for this purpose: a supplier-independent open standard for digitizing the IT value chain.

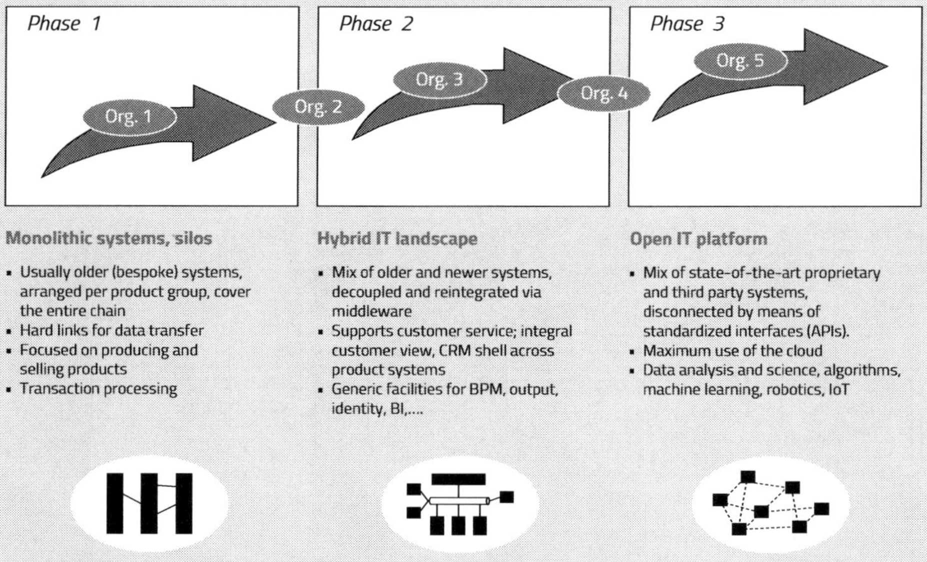

In the following chapters, the five digital breakthroughs are addressed one by one and discussed further according to the pattern in Figure 11. Depending on the situation, you will either tackle them in parallel or give priority to one or more themes. This mainly depends on what you already have in place and what will be relevant for your business in the coming period. In any case, it is always good to use an external necessity (for example, a new competitor, dissatisfied customers, price reductions in your market ...) to organize the necessary change pressure in the company.

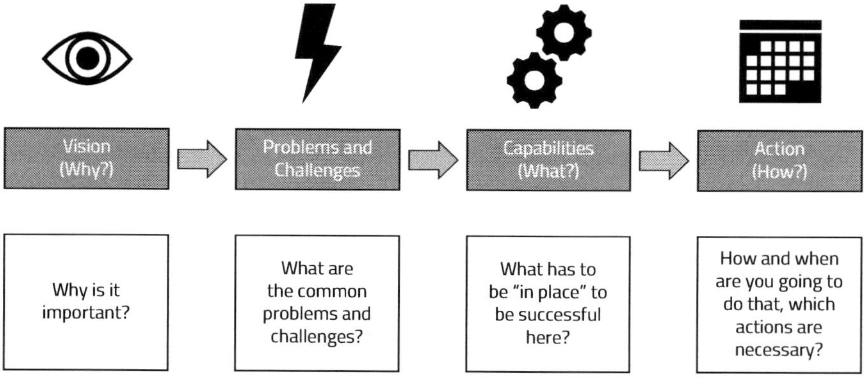

Figure 11 Approach to handling digital breakthroughs

4 Digital breakthrough 1: Data-driven organization

4.1 WHY IS DATA SO IMPORTANT?

Almost everything in our daily lives becomes measurable and used as input to recognize patterns. For example, to live healthier and smarter, by registering your sleep pattern or measuring the number of steps taken with a wearable. Other increasingly important sources of data are sensors, for example in production machines and cars, so that much more accurate preventive maintenance and replacement can take place. As a result, organizations are faced with a huge increase in volumes of data, which enables them to discover connections and trends through data analysis and data science. Technology offers more and more opportunities to make smart use of this collected data. You can get to know your customers better and better: you know where customers live, what their buying history and internet surfing behavior is, and you know what preferences they have. By analyzing the available data and combining these analyses, companies can offer targeted, even personalized propositions to customers. It is no longer necessary to use personas; data makes it possible to respond even more to the individual wishes of consumers.

All of this is also referred to as "big data". However, data is only of value if your organization knows how to deal with these enormous quantities in the right way. If, as an organization, you can ask the right questions and formulate hypotheses, you can use the "big data" to look for scenarios, the effects of decisions and answers to questions.

You can compare data with other assets of your organization such as money, people, buildings and machines. All these assets must be well managed: they are maintained, someone is responsible for them, they are secured, and an asset is replaced in time. This also applies to data. And just like other assets, data can be sold. Income can be derived from it.

> **Reading tip:**
>
> *Data Management: a gentle introduction*; Bas van Gils, 2020

Some facts and statements about data:

- There are now more than 50 billion smart connected devices worldwide, all designed to collect, analyze and share data (Forbes).
- In 2020, there were more than 6.1 billion smartphone users worldwide; 1 trillion photos were taken and billions of them shared online (Forbes).
- More than 90% of all data in the world has been generated over the past two years (Business2Community).
- We perform 40,000 searches every second (on Google alone), or 1.2 trillion per year (Bernard Marr).
- The total volume of data is doubling every year (Business2Community).
- Estimates indicate that improved healthcare, by better integrating big data, could save as much as $300 billion per year - which amounts to reducing costs by $1,000 per year per person (Bernard Marr).
- For a typical Fortune 1000 company, a mere 10% increase in data accessibility results in over $65 million in additional net revenue (Bernard Marr).
- Retailers who harness the full power of big data can increase their operating margins by as much as 60% (Bernard Marr).
- Currently, less than 5% of all data is analyzed and used (Bernard Marr).

Many leaders of major companies are now aware of these facts and the importance of data. Key data-related programs are at the top of their change agendas. They realize that a lot of money has been invested in information systems but without standardizing, modeling and cleaning up their data, they are still not getting the right information upon which to make good decisions. They have business units that use different definitions, calculate statistics differently and produce reports that are difficult to compare. The data is insufficiently reliable, not current and inconsistent. Adequate data management is crucial for organizations to achieve so that they have reliable, predictive analytics and can extract value from data; for themselves and for their stakeholders.

Using data is not only about more insight and better management and decision-making, it also goes much further than that. Roughly speaking, four different objectives can be identified with data, these are shown in Figure 12.

Insights are essential for making the right choices. To discover these insights, organizations must analyze processes, conduct continuous market research and use all available data sources. Policies and decisions will be more targeted because a broader, more complete and more objective picture is available. The transparency of decisions will be increased.

A well-organized data management foundation is important to be able to analyze scenarios and predict the impact of developments on your own organization. It will lead to greater agility; faster and better responses to changes because analyses

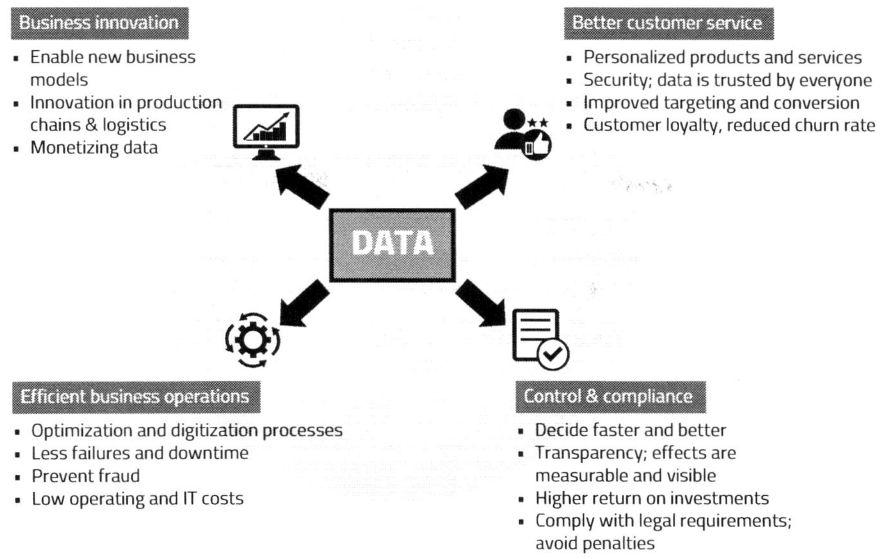

Figure 12 Data objectives

provide direct insight into what is happening and what will happen. A stable and overarching basis for data management not only enables organizations to make better decisions but also to comply with legislation, deliver higher quality services to customers and even develop new business models. Products and services can be delivered at lower costs and with shorter lead times, because data analysis is used to identify bottlenecks in processes and to further automate them where possible. An organization that knows how to transform data into usable forms can ultimately make the difference in the market.

4.2 PROBLEMS AND CHALLENGES WITH DATA

In daily practice there are quite a few problems and challenges with data that need to be recognized. These are of course different for each organization, but the following are common:

- Lots of data is not available in real time. It is there but you can't access it or it is too late when you do so. Insights from data analyses are therefore limited, not integrated and there is a need for better management information. Models and reports are descriptive and not yet predictive.
- Data management is often fragmented; the same data is captured and managed multiple times, with different definitions. Data management is largely done manually. People are not ready to manage huge amounts of data.
- Data quality (accuracy, completeness, timeliness) is not always in order; there are multiple versions of the truth. The question is whether information can be trusted sufficiently.

- There is increasing regulatory pressure from governments to protect consumers such as the General Data Protection Regulation. And the risks of cyber-attacks are increasing daily.
- Technology is outdated. Existing (customized) systems are running up against their limits due to outdated infrastructure and are no longer able to process the overload of data (e.g. due to the increased use of IoT and sensors).
- Many organizations are not ready. There is unclear ownership and control, and data competencies and skills are inadequate. There is a compartmentalized approach to improvement, the wheel is often reinvented several times, knowledge and experience are insufficiently shared.
- People are not yet sufficiently able to extract value from data. Important choices are made without quantitative support. Even if all kinds of preconditions are met, it is not yet clear what the valuable applications are and how they can be realized and used.

Is this recognizable? If so, this stands in stark contrast to the ambition to become a data-driven organization. There is certainly work to be done!

4.3 WHAT DOES A DATA-DRIVEN ORGANIZATION LOOK LIKE?

A data-driven organization is one that manages based on facts and not gut feelings. As a result, decisions can be made better and often faster. A data-driven organization enables future developments to be better predicted, so that the best possible response can be made. The goal is to gain greater insight and knowledge every day, across the full breadth of the company and the market in which it operates. We must then be able to rely on the data being of good quality.

We can look at data in many ways. There is *structured data*, organized and stored in databases. Data in the form of emails, text messages, photos, videos and sound recordings is referred to as *unstructured data*. And then, of course, there is *physical data*: archives, forms, hard-copy documents. Another cross-section concerns internal data, which we generate and manage ourselves, and external data, which we obtain from outside. If we look at the use of data, we can identify many applications. Data is the raw material for your primary business processes, for example sales, service, production, administration and purchasing. Data is presented to the outside world, to your customers and business partners, via websites and customer portals. Data is used to steer, to make the right decisions. By a manager for example, but also by a marketer for a retail organization, an actuary for an insurance company or a project developer for a municipality. We refer to this application of data as business intelligence or (advanced) analytics. Data can also be the raw material for smart algorithms and for advanced AI applications and machine learning. And data can be an independent product; data can make money.

Data science is the field of study that focuses on gaining insights from data. It is a combination of statistical knowledge, algorithms, subject matter expertise and programming skills, which are used to make connections and generate valuable insights. There are several methods for making predictions and analyzing trends. These are descriptive, predictive and prescriptive analytics.

Descriptive analytics is used to identify past trends, providing insight into what happened and why. For example, why did one marketing campaign fail and another succeed? *Predictive analytics is used to* predict what might happen. If we introduce a new product to the market, what sales can we expect? Or when should maintenance take place, using historical maintenance data combined with current data from IoT sensors? One step further is *prescriptive analytics*. This can be used to give advice on possible outcomes for various scenarios. In other words, what actions should we take to make the prediction come true? For example, if we choose the third option, what will be the effect on our customers? What do we need to do to make this work?

Example of municipalities

Data-driven organization is a spearhead at several larger municipalities. More and more data from sensors, social media data and datasets from partners and from the municipality itself are being used, always with the aim of delivering more public value to the city, fitting within laws and regulations. Examples of innovative data applications are[5]:

- Increase the understanding of the functioning of the city, by structuring and analyzing all available data on water, energy, food and material flows that pass through the city, with the aim of developing a sustainable circular economy.
- Conducting measurements of air quality, noise, vibration and meteorological indicators, aimed at addressing air pollution/nuisance and improving health.
- Developing a dashboard, which will visualize concentrations of crime in the city, with the aim of improving security in neighborhoods with a differentiated approach area by area.
- Using data to manage traffic flows. Instead of expensive management and maintenance of loops in the road surface, use social media posts from visitors, location data from cell phones and real-time data on available parking spaces in parking garages.
- Increasing insight into developments in demographics, income development, health, demand for care and the labor market, with the aim of developing more effective policies.

5 Drawn in part from an inventory by the Association of Netherlands Municipalities – VNG.

- Data analysis of housing market development, with the aim of preventing the housing market from overheating.
- Predicting the volume and nature of residents who (will) use a municipal scheme in the social domain.

In part, the above applications are used to develop policy and determine a picture for the deployment of interventions. Some data is also shared with the city, whilst other data is only set up in super secure environments.

By experimenting with and developing the various data applications, the municipalities are gradually increasing their data capabilities (knowledge, experience, processes and technology) and, therefore, the possibilities for the smart use of data are growing.

The increase in the amount of data means that such analysis methods are becoming increasingly interesting for organizations, for example, predicting purchasing behavior or anticipating movements within the commodities market.

So, what do you all need to have in place to become a data-driven organization? Figure 13 provides an overview of what is required; the next section explains the various components in more detail.

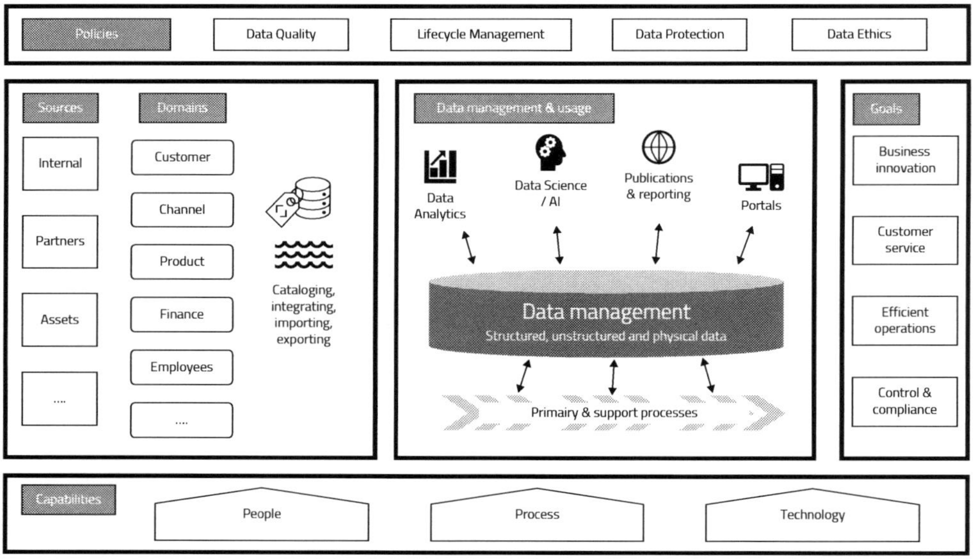

Figure 13 Data management and data usage overview

4.4 HOW TO BECOME A DATA-DRIVEN ORGANIZATION?

The challenge your organization faces is: accelerate the realization of the ambition to become a data-driven, customer-trusted organization. But how do you do that? How do you get more value from data, innovate products and services using data, make faster and better decisions, and improve operational processes with data? How do you ensure that business risks are minimized and that you comply with laws and regulations?

You don't become a data-driven organization overnight, you must organize a lot of things within your company, at all levels. You need capabilities in terms of people (skills, knowledge and behavior), process (data governance, roles, collaboration) and technology (systems for data storage, data retrieval, metadata management, tooling for data analysis and AI). Data management and quality management must be properly embedded in the primary processes of your organization and clear frameworks and ground rules must be established regarding data quality, data security and privacy. Once all of this is at a sufficient level, you can experiment, learn and scale up quickly.

Using an incremental approach, ensure that the following things are put in order, step by step (see Figure 14):

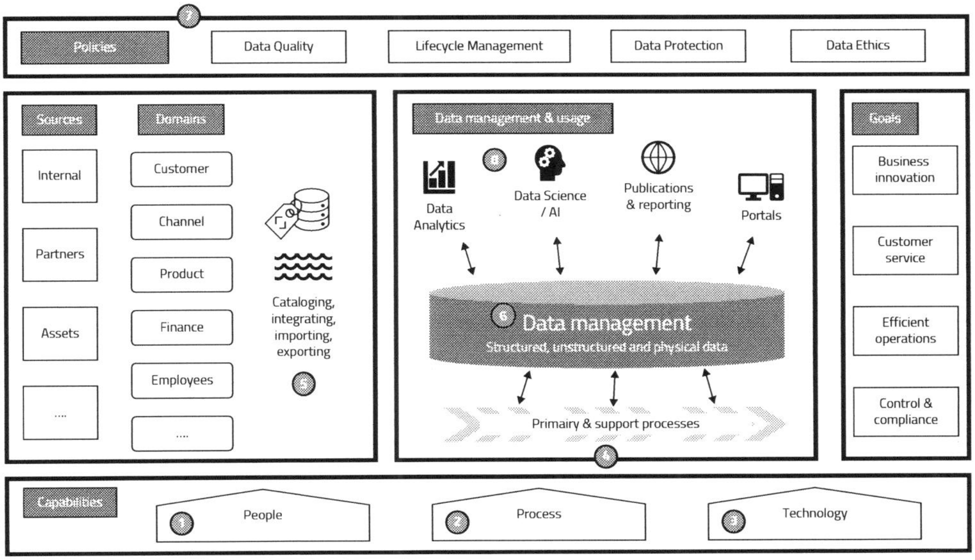

Figure 14 Action points plotted on the overview

1) Managers and employees are aware of the importance of data and have the appropriate skills, knowledge and behavior. They can ask the right questions, derived from the business goals. They remain in the lead, of course, but are much better supported in making choices and decisions.
2) Data roles such as data owner, data steward and data analyst have been worked out and invested in the organization. There is a governance structure for alignment, knowledge sharing, prioritization and escalation. There are KPIs and a dashboard has been developed to monitor status and progress.
3) IT architecture and IT solutions align with data needs. For example, a data lake, with functionality for cataloging, integrating, importing, exporting and publishing data. And a data lab, which translates the data and patterns into insights, knowledge and intelligence.
4) The organization is set up to extract maximum value from data. For example, with a "hub and spoke" model. The "hub", one central team, consists of a combination of an innovation lab, specialist competencies, working methods and technologies. The hub is strongly connected with the "spokes": various decentralized innovation teams within functional departments, business units or divisions. The hubs and spokes exchange knowledge, experiences and solutions.
5) Access to registrations has been arranged. Agreements have been made with suppliers of data (internal and external).
6) Data quality, lifecycle management and data access are continuously managed. Measuring and improving are part of the business processes. Data management is in place.
7) There are clear frameworks and ground rules about data quality (accuracy, completeness and timeliness), information security, privacy and ethical issues.
8) You have a portfolio of initiatives, to extract value from information. Derived from and prioritized to the business strategy.

You can't change the world overnight, so you need to focus. What goals are you pursuing first? If your organization is in the transition from phase 1 to phase 2, the emphasis should be on gaining control of data and you are looking for improvements in the sphere of better management, more efficient operations and better service. If you are in the transition from phase 2 to phase 3, the focus will mainly be on growth and business innovation. What kind of data is it about? Trace the authentic sources, unlock and integrate them, check the data quality and, if necessary, improve it. Take advantage of concrete opportunities or problems within the organization. Experiment and start practically with small steps: start right away, learn by doing and in this way achieve tangible results quickly. When successful, implement them more widely in the organization. Work under an architecture: each increment fits into the situation to be achieved and forms a part of it as a "puzzle piece". In this way you can build, on a step–by-step basis, the foundation of a data-driven organization.

Example of an insurer

An international insurance company conducted a study of possible use cases of data-driven organization:

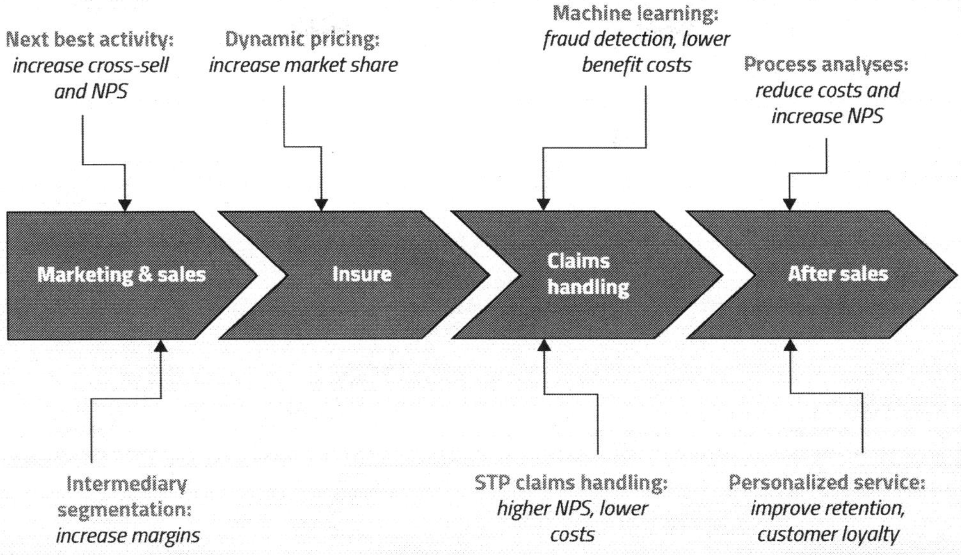

A short business case for each use case was prepared. Based on this, a decision was made to develop a data ecosystem consisting of the following components:

- Data lake: focused on the collection, classification and storage of large amounts of structured and unstructured data.
- Import and export hub: facilities to securely import and export data to and from third parties, the market, regulators, etc.
- BI portal: analysis and reporting functionality, focused on providing insight into past (what-was), present (what-is) and future (what-if).
- Data lab: intelligent playground to experiment, build new applications and train people, using data, patterns and AI.
- Omnichannel hub: focused on unlocking back-office systems for customer interaction across all channels (customer portal, intermediary portal, contact center).

Because the independently operating subsidiaries are all in different phases and have different priorities, the data ecosystem is used slightly differently everywhere: in one to comply with new legislation (IFRS), in another for risk management and fraud detection, and in a third unit as a tool to provide customer portals with real-time information about customers and policies.

5 Digital breakthrough 2: Smart, digital processes

5.1 WHY ARE SMART, DIGITAL PROCESSES SO IMPORTANT?

In Chapter 4 we set our sights on data, now we shift it to processes. Processes also have a major impact on the ins and outs of your business. And here, too, a digital breakthrough is needed. So, it's not surprising that many organizations feel the need to improve, tighten or even completely revamp their processes. The better the process, the more efficient and effective the work can be. Designing processes intelligently and digitizing them makes human intervention less necessary and will lead to a substantial simplification of business operations in addition to structural cost savings. Digital processes can ensure that products and services are delivered flawlessly, correctly and quickly, and that creativity and craftsmanship are used where they provide real added value. And they make remote working possible, for the organization's own employees, and with customers and business partners.

New technological developments offer opportunities to radically improve processes. Machines are becoming increasingly intelligent and can take over human tasks ever more easily and quickly. They can ensure shorter turnaround times, lower stock levels and fewer errors. New technologies make working (together) easier. And you are less vulnerable to scarce expertise. The worlds of industrial and administrative processes are increasingly being integrated, for example the automatic ordering of spare parts based on continuous measurement of the maintenance condition of a machine.

Satisfied customers, low operating costs and high employee satisfaction are the outcomes of smartly designed, digital business processes. The key questions are:

- How can we radically simplify our business operations and make processes more efficient, cheaper and of a higher quality?

- How do we ensure that we digitize processes as much as possible and deploy people to those activities that really matter?
- What added value do new technologies such as dynamic case management, robotic process automation, artificial intelligence, social technologies and process mining offer?

The challenge your organization faces: (re)design, streamline and digitize business processes, where advanced technologies dramatically improve performance.

Reading tip:

Digital Transformation; Jo Caudron and Dado van Peteghem, 2018

5.2 PROBLEMS AND CHALLENGES WITH SMART, DIGITAL PROCESSES

Process management has - fortunately - been high on the agenda of many organizations for years, thanks in part to movements such as Lean and Six Sigma, a collection of quality management methods with which one can improve the quality of business processes step-by-step. The origin of these methods comes from the manufacturing industry, including such as Toyota and Motorola, but they are now also widely used in the more administrative and information-intensive environments. Despite this, there are still quite a few bottlenecks that can be experienced in terms of how processes are handled:

- **Unclear processes:** In many organizations, it is not very clear how processes work. In such cases they are difficult to organize, and you do not know whether these processes function effectively. You do not know whether the work is done properly, and you have little grip on the progress. If it is not clear how processes run, then implementing changes becomes almost impossible. Manuals with process descriptions are often drawn up, not least under pressure from the accountant or the internal audit department, or because there is an ambition to obtain a quality stamp (ISO 900x). But after a while these manuals and the actual reality no longer match, and they disappear into a deep drawer.

- **Data quality not in order**: Data is an important input for processes. As we have already seen in Chapter 3, if data is of insufficient quality, contains errors, is not up-to-date, and contains contradictory information, things will inevitably go wrong in the process and the result will not be correct: "garbage in = garbage out". For example, due to the poor quality of the supplied information, incomplete documents enter the process, resulting in a lot of extra work, downtime and even errors. If you allow incomplete invoices, application forms or files in the process, the processing will be delayed, and time-consuming workarounds will be needed to handle these incomplete matters.

- **Too much complexity in processes**: Processes, as a rule, are not designed but created; they have grown organically over time. As a result of specific customer requests or rare events, process steps have been added. Customization has crept into processes that could be organized in a standard way. Exceptions confirm the rule.

- **Insufficient management information**: Measuring is knowing. And the chain is as strong as the weakest link. Where are queues, accumulations of stock, wastages, problems with delivering on time? If you do not have sharp management information, feedback from customers, business partners and employees, then you cannot learn, intervene and improve processes.

- **Dependence on scarce knowledge**: It is not unusual for an organization to become dependent on specialists with scarce knowledge when executing processes. Specialists who, as a rule, do their best but who (often unconsciously) ensure that they become indispensable, and that the transparency of the organization decreases even further. Salary demands are almost unavoidable and restructuring and changing becomes a very difficult, if not impossible, task. If these specialists move away (for example, transfer to the competitor or retire), no one can absorb it and the organization's survival is seriously disrupted or sometimes even threatened.

- **Difficult cooperation**: Processes don't care about departmental boundaries; they run right through your organization, even beyond it. Different departments and business partners carry out different parts of the process. However, particularly in functionally structured organizations, you often see that cooperation is not always optimal. Employees supervise their own field of work, but do not usually oversee the links in the neighboring process, and certainly not the entire business process. For example, it may not be clear to many employees who is responsible for authorizing a payment. The result is that the finance department then experiences problems. Or the logistics department cannot find out from the receipt to which location the goods must be delivered. An us/them culture develops, and people point at each other when things go wrong.

- **Digitize existing processes**: Process improvement is often mentioned in the same breath as digitization. But digitizing existing processes does not automatically mean that they will also improve. In fact, you run the risk of a historically developed situation being "cast in concrete". It gets even worse when you use a standard IT solution. For example, you have purchased an ERP package to replace the current custom-made systems. If you adapt this ERP solution to your processes, instead of adapting your processes to the best-practices that are contained in this standard solution, you have developed the legacy of the future. The cause often lies in resistance to change. You typically hear "this is how we have been doing it here for years, this change does not work, we have tried it many times". Instead of radically reducing complexity you have created new complexity.

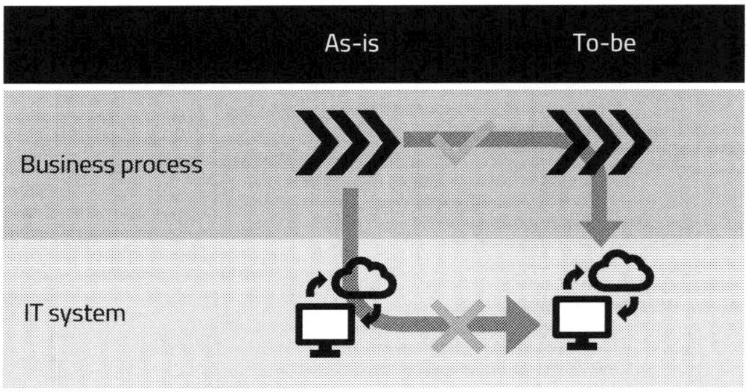

■ 5.3 WHAT DOES IT TAKE FOR SMART, DIGITAL PROCESSES?

A process is a chain of activities that takes place between the demand for a product or service and its delivery. This series of activities is carried out by people, computers and/or production machines and can cut across different departments, and sometimes even different organizations.

The start to achieving a smart digital business process is understanding the business process itself. Where does the process begin and what is the concrete reason for starting the process? What product or service is delivered and what activities are necessary to produce it with the desired quality? Insight into the business process is obtained by making a design of it, usually in the form of a process model. Depending on the situation this can be done bottom-up, by mapping the current process and making improvements from there, or top-down by making a redesign for the desired situation. The rule is: make sure you have a good process in place before you start using (new) technology, do not start digitizing existing business processes blindly. But do look at the possibilities that digital technology offers, from improving processes though to the creation of new process models, thereby completely replacing existing processes.

When improving processes, use is often made of the knowledge and experience built up in Lean management[6]. This involves a structured way of identifying and eliminating waste, by making a distinction between activities that add value for customers and activities that do not. Subsequently, improvements are determined and implemented, and the effects of these improvements are made measurable.

6 James P. Womack and Daniel T. Jones, "Lean Thinking: Banish waste and create wealth in your corporation", June 2003.

Principles of Lean management

Lean is a management philosophy that aims to eliminate waste - activities and resources that do not add value: the goals include a reduction in lead time ("from sand to customer"), increased production quality and reduced production costs. Lean manufacturing was largely developed by Toyota's founder, Sakichi Toyoda. By applying Lean manufacturing, Toyota was able to fundamentally reduce costs and grow into one of the largest companies in the world. Because of this success, the philosophy has been widely applied by other organizations. Lean management principles build on those of Lean manufacturing; initially these principles were set up for production processes in the manufacturing industry, but they can also be applied to administrative business processes:

Reading tip:

Toyota Production System; Taiichi Ohno, 1988

1. Overproduction: this is considered the worst form of waste. Production takes place when no customer has asked for it. It occurs when processes continue to work when they should stop.
2. Waiting: this is caused by a sub-process or supplier not delivering on time. The waiting time further down the process is then often filled with activities that do not add value or, worse, it can lead to overproduction.
3. Transportation: this involves the unnecessary movement of materials and semi-finished products such as work-in-progress from one operation to another.
4. Redundant process steps: many process steps are unnecessary and could have been avoided. Think of repair work, reprocessing, handling and storage resulting from defects, overproduction and buffers that are too large or too small.
5. Stock: this refers to storage that is not directly required for current customer orders. This includes raw materials, spare parts and semi-finished products as well as finished products.
6. Unnecessary movements: these are made by people or machines, because work is done in a poor layout or because the workplace is not set up properly. For example, there is too much walking required to pick-up the necessary materials and tools.
7. Defects: these are products or services that do not meet the requirements or expectations of your customers. Defects can arise, for example, from working with incorrect information. This leads to returns, complaints and sales that cannot be booked as turnover.
8. Talent: this is a waste that was added later and occurs when the knowledge and skills of employees are underutilized.

There are all kinds of new digital technologies that can help to improve and renew processes. Technology can play different roles in this. In addition to automating manual activities and decisions, one can also look at integrating activities so that there are fewer transfer moments in the process. Transfer moments often lead to longer processing times, duplication of work and require time for coordination. But shifting activities to customers and chain partners, performing activities in parallel and bridging distances are also possible with new technology. Making (specialized) knowledge accessible can result in processes becoming more efficient and sometimes make it possible to skip intermediate links. The control of the process can be improved by using technology that makes the course of the process transparent and predictable.

Without pretending to be exhaustive, the following is an overview of the most important technologies and best practices. Reality is less structured than the list suggests; the technologies and concepts have arisen from all sorts of different angles, sometimes converging, sometimes overlapping. Various technologies can be combined, and new forms of applications emerge.

1. **Straight through processing (STP):** This means that an order is processed directly without human intervention. The principle originates from the banking world but can be applied anywhere. Business processes that are set up based on STP are executed in real-time and automatically from the first step (after the start by a trigger). Only exceptions within the process may still be processed manually. The main advantages of STP are a shortening in the turnaround time and a reduction in errors. Data input, often the trigger for a process, is shifted to customers, chain partners or devices by means of portals, web services and APIs for data exchange. Input controls and product rules occur as early as possible in the process, so the factory (or back office) receives "clean orders" and production processes are not disrupted by incomplete or incorrect data. See Figure 15 for an example: at least 90% of all customer requests are received in a structured way via digital channels. By controlling the input at source, failures are "kept out of the factory", resolved with the customer and measures are taken to reduce the chance of failures in the future. Customer requests are STP processed in this way, whilst the output to the customer is ideally in digital form.

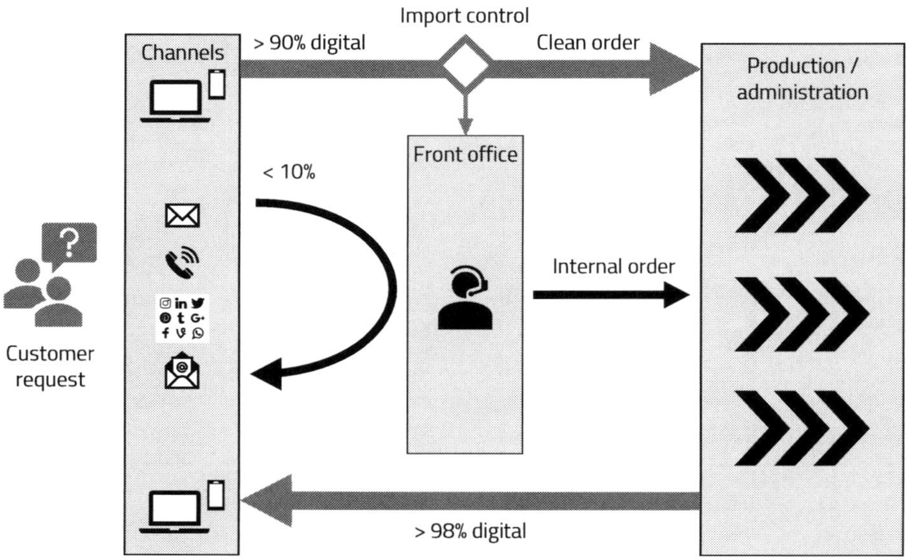

Figure 15 Example of straight through processing

2. **Social technologies:** A collective term for technologies that enable organizations to improve internal and external communication, enhance collaboration and promote knowledge sharing. This could include technologies such as blogs, wikis, groupware or collaborative software and social networks. It enables location- and time-independent work and can increase the reach and attention of the organization, strengthen knowledge and provide better access to information.

3. **Process management software:** This is a collective term for solutions used to map, evaluate and optimize structures, processes and workflows within a company. Several variants are available: workflow management, business process management (BPM) and (dynamic) case management. Their use depends mainly on the degree of complexity and changeability of the process. Think of the difference between handling a driver's license renewal (simple process) versus handling a seriously ill patient (complex process). The more complex, the more options for routing and process handling must be available. Basically, these solutions ensure that the work goes from one employee or department to another in an automated and optimal way. In doing so, the information, which is already available in the organization, is utilized to the maximum. This information can be retrieved from various applications without requiring an in-depth knowledge of the different systems. It provides insight and control over processes and supports continuous process improvement.

4. **Process mining:** The umbrella term for a collection of methods, techniques and tools that you can use to visualize, expose, analyze, monitor and improve the actual course of business processes. This enables you to screen the actual process for bottlenecks, waste, deviations and contradictions compared to the initial design of the process. With the help of process mining, you can discover how your business processes performed and you are able to detect and solve any hidden costs, inefficiencies and bottlenecks in processes.

5. **Algorithms and automating decisions**: A business rule engine can be used to support decision-making. How much discount can you give to a particular customer? Is someone eligible for a benefit? Can you pay an insurance claim or not? Business rules (such as sales terms, price tables and product specifications), laws and regulations and smart algorithms are all captured in a business rule engine and executed automatically. These rules are not part of an application but are managed separately by users who are most affected by the rules: doctors manage the rules for clinical decision-making, lawyers manage the rules for legal advice, and actuaries manage the rules for reviewing insurance claims.

6. **Process automation:** The use of robotic process automation (RPA) or physical robots can take over manual work performed by employees. RPA is the use of software robots (also known as "bots") that can interpret and automate existing

work processes. Automation allows repetitive business processes to be optimized and standardized. It provides the opportunity for organizations to scale up, perform activities faster and minimize errors. An example of this is the chatbot: this is an automated solution that allows you to answer customer and employee questions in a chat conversation. The chatbot is "trained" to answer these questions and works with pattern recognition; to each pattern an action (usually giving an answer) is linked.

7. **Artificial intelligence (AI):** This involves computers performing tasks that would normally require human intelligence. Many applications have long been integrated into our daily lives, such as search engines, spam filters, chess computers and robot lawn mowers. Increasingly there is also face and voice recognition, pattern detection and research. In the future, this could grow into a form of intelligence that can reason and solve problems. The potential applications for AI are endless. Many applications are based on algorithms: a set of instructions for solving a specific problem. This could be performing calculations, processing data, or performing tasks automatically. Machine learning is a subset of AI and involves algorithms that can learn and make predictions based on data, where the algorithm can adapt and improve itself.

8. **Digital twin:** A digital twin is a virtual model that represents a physical object, with the purpose of understanding its status, responding to changes, improving and adding value. In industry, the concept of digital twins has existed for some time. Using sensors and computer models, wind turbines, buildings, bridges and planes are digitally simulated to discover how they perform in different conditions. Digital twins are becoming more intelligent, more robust and larger. The concept is now also applicable to organizations. Digital twins of a department, a business unit and even an entire organization are possible. It is a real-time model of the organization, in line with its strategic goals. It monitors and analyzes how people work in processes, with production tools and workflows between departments. With a digital twin of the organization (DTO), you can play out scenarios, analyze them, optimize them and experiment with them before adjusting things in practice.

9. **Blockchain technology:** A blockchain is a database of trusted transactions, copied to all parties involved in a business process. Each activity in the process is a transaction that, after validation, without the intervention of a third party, is recorded as a "cube" on the "chain". In this way the process is controlled via interaction with the blockchain exactly as it was intended: participants in the process can only work as agreed beforehand. These agreements are recorded in a process model. This creates grip and control over these business processes. For the time being, blockchain will mainly offer added value in chains that transcend companies. Existing applications include crypto money (such as

bitcoin), financial transactions, passing of property deeds, signing of contracts and transfer of medical data between hospitals. Blockchain technology can also be used in making value chains both transparent and automated. This involves various actors in a value chain placing their information on a blockchain, which makes the progress of the process transparent or reveals the origin of a product.

10. **The internet of things (IoT):** This consists of physical objects such as cars, home appliances and wearables, which are connected to the internet through sensors and can send data online. Well-known examples are movement and fitness trackers, health meters, tracking & tracing of means of transport, smart thermostats and security systems. The growth is enormous, it is expected that tens of billions of objects will be part of this network in the coming years. From smart watches, toys and bicycles to medical equipment, bridges, trains and equipment in agriculture and livestock. The possibilities are almost endless. Business assets and end products independently produce maintenance reports and perform diagnostics. They detect malfunctions, report them online to technical support and suggest possible solutions themselves, from ordering spare parts to replacing them with a completely new device. The internet of things is bringing about enormous changes in how processes run, eliminating intermediate steps and providing completely new insights into process performance.

11. **3D printing:** This allows for the efficient production of unique products where and when they are needed. Through prototyping and research and development, cycles can be greatly shortened. 3D printing can be used to create spare parts, without the large inventories and logistics required to distribute them. 3D printing makes it possible to create personalized products tailored to customer needs and preferences. Although we associate 3D printing with plastics, the number of materials that can be 3D printed is growing rapidly. From recycled plastic to metal and from tissue to chocolate.

12. **Human augmentation:** This refers to the collection of technologies that improve human productivity or capabilities and that contribute in some way to the human body. Human augmentation allows people to see, hear, feel, and lift many more pounds (by adding an exoskeleton), among other things. It offers the opportunity for organizations to deploy technologies where humans themselves fall short. This allows certain internal processes to be more efficient.

The above overview is by no means complete and will be outdated tomorrow, since technological developments take place at a rapid pace. However, all are striving for the same goal: excellent processes, cleverly designed, supported by the right technologies. And therefore, able to deliver products and services flawlessly, right the first time and tailored to customer needs.

5.4 HOW TO BECOME A DIGITAL, PROCESS-ORIENTED ORGANIZATION?

In the transformation to smart, digital business processes, several steps are important. If it hasn't already happened, it starts with understanding current processes. Knowing what processes run through your organization, who adds what value and what waste, bottlenecks and improvement potential can be identified. Then (re) design the business process to get rid of existing inefficiencies and other limitations. Formulate the goals and principles of the process in terms of quality, lead time, costs and compliance with laws and regulations. Derive KPIs from the strategy and make it possible to measure what is important and show whether you are on track. Ensure that tools aimed at quality control of processes are in order: process documentation, procedures and standards, periodic reviews and management reports on numbers, lead times, dropouts and capacity utilization.

Business processes also require well-organized management to steer in a balanced way towards goal realization, innovation and risk management. Appoint process owners with a mandate, who steer - right through the organization - to the realization of the formulated objectives. The process owner facilitates those involved in the process and involves employees in process improvements and process innovations.

The newly designed business process (and therefore not the existing process) is the basis for making choices on the technologies to be deployed and finding the appropriate solutions and suppliers for these. This is an interaction: based on new technologies, you discover new possibilities that can lead to further modification of the process design or even to radical innovation. Invest in knowledge and gain experience with new technologies. Delve into this actively, research what makes other companies successful and which technology can be applied where. Be curious, experiment and show courage.

It is wise to start small with a proof-of-concept, prototype or digital twin. Mimic the optimized process design, see what the effects are and adjust things based on the insights obtained. Only when the results are sufficiently satisfactory should you start scaling up and implementing more broadly. Evaluate the processes continuously, measure against the formulated KPIs and adjust where necessary.

Digital processes and design thinking

The topic of smart, digital processes is ideally suited for the application of design thinking: a method with which you generate alternative scenarios and solutions in co-creation, focusing on the needs of stakeholders. Design thinking in combination with Lean management combines out-of-the-box thinking with the realization of efficiency and innovations in business processes.

Reading tip:
The Design of Business; Roger L. Martin, 2009

Design thinking is a creative approach to problem solving, seeking to balance analytical and intuitive thinking. Many managers and analysts are not particularly strong in creative thinking, whilst on the other hand designers often have difficulty with analytical and structured thinking. If both approaches are brought together when looking at innovations, new and surprising insights often emerge. An important factor in design thinking is that the human being and his needs are central in the solution. Stanford University has developed a method known as the "five phases of design thinking":

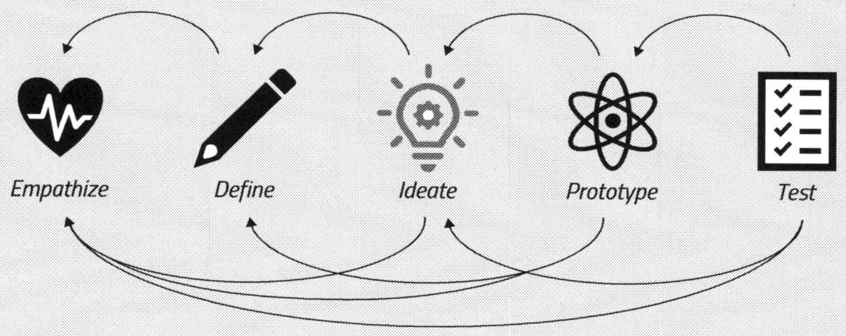

1. **Empathize:** this involves obtaining a good understanding of the problem. Dive into the problem: look at the history, context, goals and who is involved. Get to know the user/target audience: their habits, values, wants, needs, desires and fears. Combine and enrich the information gathered. Make outcomes understandable and inspiring.
2. **Define:** based on different perspectives, you will then work with a team to determine the problem or challenge: what is going on? Dig deeper and find patterns. Define the problem you want to solve from the user's perspective, agree on the design challenge.
3. **Ideate:** this is about coming up with as many solutions to the problem as possible with the team that are "outside-the-box". There are all kinds of techniques for this: brainstorming, brainwriting, mind mapping, "worst possible idea", role playing etc.
4. **Prototype:** in this phase you present the solutions to others. Try things out. Make (simple) prototypes; failure is not so costly at this stage. Try to get as many reactions as possible to potential solutions in order to further refine ideas. This will give insight into which solution works and which does not.
5. **Test:** try out the prototype with users, experts and stakeholders. Use their feedback and test results to improve the solution.

The five-stage model appears to be a linear process, but it is not. In the process, for example, different researchers and teams can be deployed in parallel on different phases. The test phase often provides a lot of input for earlier phases; you generate new insights, as a result of which the prototype has to be adapted. Or a team may discover at a later stage that the problem has not been properly defined, causing you to "fall back" to the definition phase.

6 Digital breakthrough 3: Brilliant customer experience

6.1 WHY IS A BRILLIANT CUSTOMER EXPERIENCE SO IMPORTANT?

Because digital technology is evolving so rapidly, the ability of organizations to be customer-centric is also changing. Disruptive phase 3 companies are creating new values for customers by applying these digital technologies in smart and effective ways. What all these companies have in common is their passion to put customers fully at the center of everything they do. As a result, not only are customers very happy to do business with these companies but they also have the same expectations of your organization. Key words are convenience, fast, best price, personal, involved and trusted. Customers expect to be helped anywhere and anytime, to have insight into their own data, the status of current transactions and contact history. They need to be served with customization and empathy, tailored to their needs at that moment. It takes a digital breakthrough to achieve a brilliant customer experience.

Brilliant customer experience is not about the next app, the latest marketing techniques or about what is possible with customer data. It's about the transition from a customer relationship that just exists, to a relationship that you really build with your customers. Digitalization and IT solutions (such as advanced CRM systems) are a precondition for this and offer new opportunities; they are never a goal in themselves. But they are indispensable in transforming into an organization that is maximally personal and relevant.

Customers will gain and maintain trust in your organization and your solutions, show greater loyalty, be more open to new products and services, recommend your organization to others or even actively refer to it. Achieving a brilliant customer experience will lead to growth and more profitability for your organization through increased sales, more cross/deep/up-sell and lower cost per transaction.

Delivering the right customer experience is a "business capability", an ability consisting of a combination of knowledge, competencies, techniques, processes and technology. You can build up this capability and if you succeed in doing so, you can make the difference in the market and take a lead on the competition. Questions that arise include:

- How do you optimally connect to the needs of your customers?
- How do you improve the customer experience and encourage engagement?
- Can you achieve a seamless customer experience across all channels?
- How do you get customers to recommend your organization?
- How do you change customer processes, employee competencies and IT to make all this possible?

The challenge you face: realizing an exceptional customer experience as the basis for a distinctive customer relationship and strong customer loyalty.

6.2 PROBLEMS AND CHALLENGES WITH CUSTOMER EXPERIENCE

Reading tip:

Service Excellence; Jean-Pierre Thomassen and Eric de Haan, 2016

A lot of mountains have been moved in recent years in terms of improving the interaction between companies and its customers. Many organizations have started, or even reasonably completed, the movement from product orientation (phase 1) to customer orientation (phase 2). Customer contact centers have been optimized, the connection between telephony and the online channel has been made. But in the current practice there's also much that goes wrong:

- Customers regularly experience the fact that their interests are not paramount in the way they are dealt with. The accessibility and handling time of customer requests are not always in line with expectations. For example, a customer might be approached by different departments separately, when they do not know about each other, sometimes with the same or even opposite questions. Several times the customer must pass on the same information, when this information is already available somewhere in the company. It happens that a deceased customer must die "three times", so to speak, before everything is properly processed in all information systems. In the meantime, their next of kin receive cheerful mail in connection with some lighthearted marketing campaign.
- There is information loss when the customer switches from one channel to another within the same organization. The customer must explain his problem repeatedly. The status of a customer request is not always known, the customer then has no idea where he stands.

- What you frequently still see is that internal business processes are playing a leading role in providing service to customers. A customer question is experienced as disruptive, consequently communication is handled by product technology. The administrative processes of the organization determine the customer interaction; customers are not helped from a customer case perspective but instead from a production process viewpoint. The result is that response rates to customer requests are too low, appointments are not properly met and customers themselves must find answers. The control over the customer, the customer data and the customer contact are all missing.
- An additional point is that customer interaction is often seen as a cost item, which results in shorter handling times and lower costs per customer contact. However, customer contacts are a "moment of truth" and offer a unique opportunity for an organization to make the difference, by responding to the (often unspoken) needs of the customer and exceeding his expectations. Customer contacts are, therefore, very decisive in the relationship that is built with customers. Incidentally, customer interaction processes are not always efficient and can lead to excessive costs. But it can all be organized smarter.
- Another common problem is that the execution power in the implementation of fundamental, customer-driven changes is too low. Information systems with customer information are outdated in terms of functionality and technology, not future-proofed and unable to adapt to the speed required by the business (and the market). Furthermore, customer data is not always of the right quality, an integral customer view is incomplete or even missing.

Customer focus is increasingly seen as a hygiene factor, a dis-satisfier. If you don't have it in order, you will lose your customers in no time. But even if you have solved the above points, you are still not there. To move on to phase 3, the new world, you need more: a brilliant customer experience.

Example of a health insurance company
A customer calls his health insurance company and wants to know more about the costs and reimbursements of artificial insemination. The hospital has indicated that there is a co-payment of €200 per treatment. However, the employee of the health insurance company thinks that six treatments are fully reimbursed but verifies this to be sure. The employee feels responsible for the client and does not rest until the client has been properly helped. He is extremely energetic, enthusiastic and friendly and manages to establish a good rapport with the customer. However, he discovers that there is conflicting information in the sources he consults: On their own website: "no co-payment". On the hospital website: "€ 200- co-payment per time". And in the internal knowledge base: "total own contribution € 200". The employee consults with his colleagues several times (with a direct colleague, a colleague from the "product development" team and a colleague from the "claims" team), and finally the claims department decides. All this while the customer is on hold. The employee shares the

news and advises the customer that the personal contribution may be deductible as "special medical expenses". The customer is very satisfied after all, as evidenced by the online survey that was filled out by this customer after this conversation. Despite this positive outcome, there are quite a few improvements that can be identified:

- Unnecessary ("waste") call: clear information on the website could have prevented the call.
- No feedback to the product/content owner: nothing from the search and findings is recorded so details do not reach the person responsible for information regarding fees. Colleagues will run into the same problem again with any similar new query.
- No content contact registration: the response to the customer is not recorded. If the customer calls again about this, a colleague will not know what was discussed previously.
- The case is particularly instructive for colleagues but is not used for the purpose of coaching and training.

6.3 WHAT DO YOU NEED TO ACHIEVE A BRILLIANT CUSTOMER EXPERIENCE?

The concept of customer experience (CX) concerns all the interactions, expectations and emotions that a customer has with an organization, throughout the entire customer journey. Customer experience is, therefore, primarily about the relationship between the customer and your organization. Customer experience says something about how the customer evaluates this relationship. Achieving the right customer experience will lead to mutual added value. To deliver this value, it is necessary that the organization has a good understanding of itself, its customers and their expectations. You will need to realize capabilities around the following elements:

- **Customer needs and life events are leading**
  You connect maximally to the experience of your customers by offering products and services along "life events", tailored to the needs of the customer. First, make sure you meet the basic needs of customers.

  Customer needs are often related to life events: changes in their personal situation or events in their daily life. Examples of events include going to school, graduating, going to work, moving to a new house, getting married, having children, getting divorced, going on vacation, retiring, getting sick (and getting better) and so on. This applies not only to the Business-to-Consumer market, but also in the Business-to-Business market where other events can be identified, such as founding, merging, moving, new management, new strategy, etc. All these events affect the needs of the customer, so it is good to know them in order that you can respond to them.

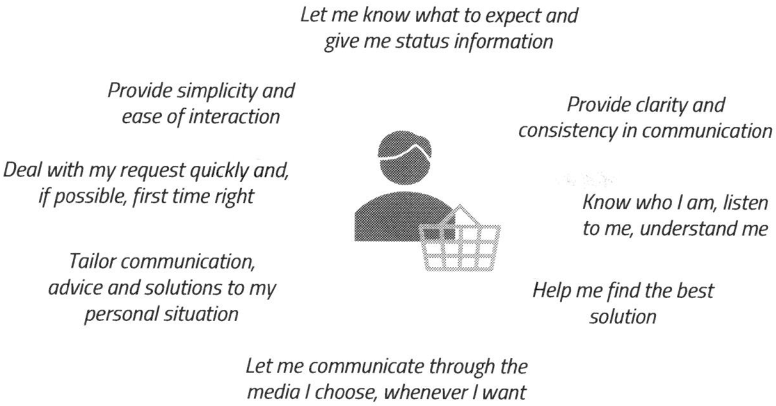

Figure 16 Basic needs of customers

- **Connecting to the customer journey**
  The customer journey is the "journey" that the customer takes: it starts with orientation, the first contact with your company until (far) after the purchase of the product or service. By mapping the customer journey, you gain insight into all touch points, the moments within the process in which you as a company meet the customer. You see which steps a customer takes before he makes a purchase from you, but also the steps afterwards, which often lead to him remaining a customer or leaving. With these insights you can introduce improvements in order to make the customer journey easier and more enjoyable. You can make sure that the customer journey is an enjoyable and fun experience, so that you get more converted visitors and are able to really bind customers to you.

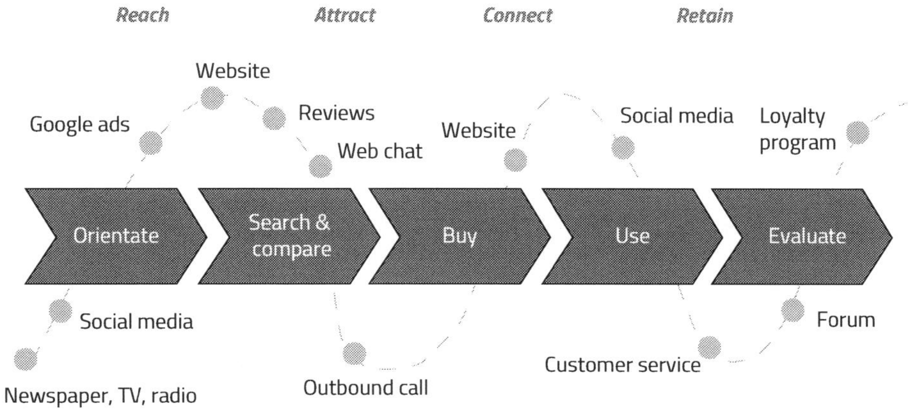

Figure 17 The customer journey and examples of touchpoints

- **Omnichannel is the new norm**
  Of course, you want an omnichannel service for your customers: the customer chooses which media (web, phone, chat, email ...) he uses to communicate with your organization. But you can "seduce" the customer to use the most efficient channels, especially online. This offers great advantages not only for your

organization but also for the customer. The online channel allows STP processing, delivering a shorter turnaround time and a lower cost, which you can also pass on in the selling price. And better accessibility (7x24 hours), faster processing and less chance of errors. But when it comes to complex and, for the customer, radical and/or emotional situations (for example, in the customer event "moving to a new house" where the customer needs a new mortgage), human interaction, a listening ear and an understanding of the issues can make the real difference.

Multi / cross / omnichannel

The terms multichannel, cross-channel and omnichannel are often mixed up. Multichannel means that the customer can choose between (usually a limited number of) channels that are not, or barely, attuned to each other. The organization makes the choice for the customer as to which channel best suits his request. In cross-channel the customer can use multiple channels, now they are aligned but the company still decides for the customer which channel is used and when. Omnichannel describes the situation where the customer chooses the channel himself, and the channels connect seamlessly.

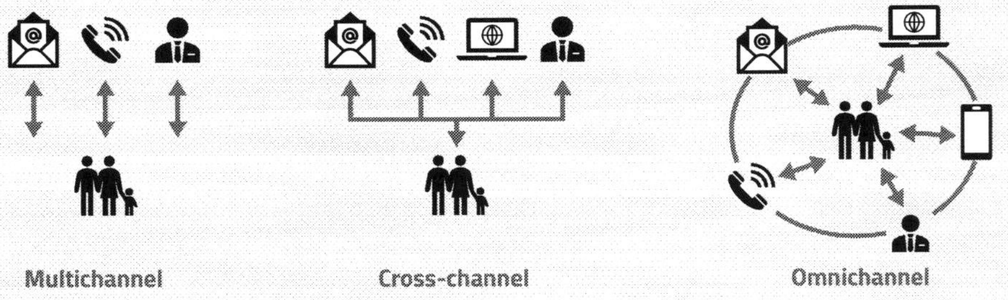

- **Proactive service delivery based on customer data**

  In Chapter 4 we have already discussed the enormous value of data. Good quality customer data is indispensable in realizing a brilliant customer experience and a proactive relationship with the customer. Based on analyses of customer data, one can map out a customer journey with touchpoints and develop a 360° customer view with all kinds of relevant characteristics. If, for example, data shows that customers who visit your web shop between 19:00 and 20:00 are mainly women between 25 and 40 years old, you can use this as the basis for an effective marketing campaign. With a 360° customer view, in combination with external marketing data, you understand the customer, his needs, his behavior and his context. It shows immediately what the customer has purchased, which customer contacts have taken place and which issues are still open. Based on an understanding of the customer, proactive action can be taken in customer interaction (next best action) or a proactive offer can be made (next best offer).

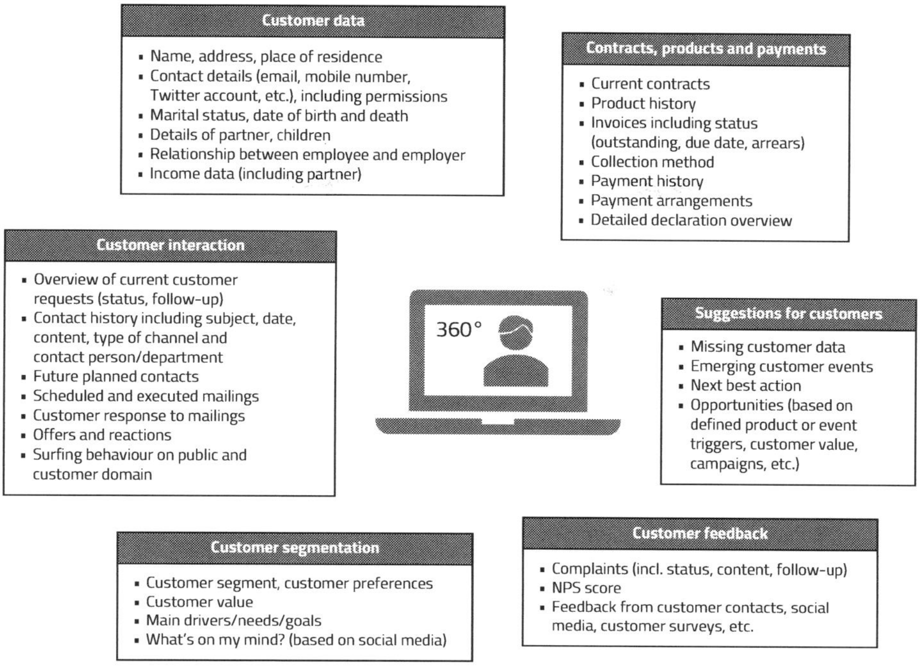

Figure 18 Example of integrated customer view

- **Maintaining a grip on customer processes**
  Customer requests are processed immediately whenever possible ("first time fix") but this is not always achievable. Additional information is often needed, for example, to process a divorce, death, claim or pension application at an insurer. These types of requests are then passed on to the relevant departments (and there can be several, especially if the customer has purchased multiple products). This is where backlogs often occur. In these kinds of situations, it is important to manage customer cases: customer requests that you cannot handle immediately are "deployed" and managed as internal orders. The aim is to monitor the progress in the process, monitor actions, report and generate status information.

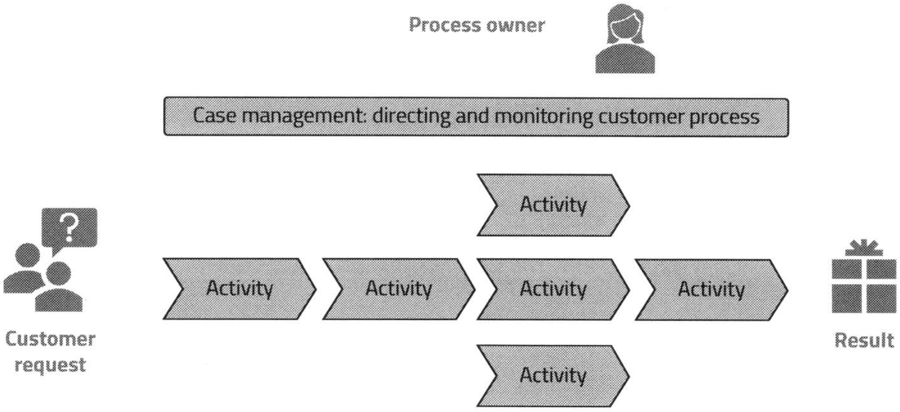

Figure 19 Controlling customer processes

As already mentioned in Chapter 5, it is important to assign ownership for processes unambiguously in the organization. A customer process is a process that leads to a result (product or service) with added value for the customer. It starts as a result of a trigger, usually a customer request, and runs through to a result, has one or more contact moments and consists of an ordered series of activities during which the execution takes place. The customer process owner is responsible for the performance of the process and ensures that it is delivered according to customer expectations. He knows what the customer expects of your organization, controls the entire chain of activities between customer demand and customer result, makes internal agreements with those involved about quality, timelines and accessibility, focuses on the achievement of Key Performance Indicators (KPIs) and ensures a smooth flow of the process through all departments. In this way, customer processes are continuously improved based on customer feedback and insights from KPIs.

Key Performance Indicators (KPIs)

There are many KPIs available to measure the success of a brilliant customer experience, for example:

- Customer satisfaction: what do customers think of your services, how satisfied are they?
- Net promoter score (NPS): how loyal is a customer?
- Customer lifetime value (CLV): how much turnover does a customer generate in the period that he is a customer?
- Customer effort score (CES): how much effort does it take to get a question resolved or a process completed?
- Customer referral value (CRV): how many recommendations does a customer make?
- First time right: how many customer requests are resolved immediately?
- Self-service%: how many of the customer requests are solved by the customer themselves through the online channel?
- Customer equity (CE): a type of average customer value, a combination of various KPIs.

6.4 HOW DO YOU DEVELOP A BRILLIANT CUSTOMER EXPERIENCE?

Realizing a brilliant customer experience is a profound process. It affects almost the entire organization and means a turnaround in all kinds of areas. You will have to work along five angles:

1. **Develop a sharp customer experience strategy:** This should fit the business strategy and brand values. Put the customer at the center, make sure the customer has control over his own situation and can take the wheel himself via the online channel. Understand that the customer is not interested in how your business is organized internally. With the right customer experience strategy, you can realize your customer goals: increased customer satisfaction, better products and services, greater loyalty and more sales. Elements of the customer experience strategy will be:

   - Omnichannel service: setting up seamless customer service across different channels. Ensure that each channel has all of the relevant information about the customer, so that the customer's personal situation and needs are known. Develop the customer relationship by creating relevant touch points through the most appropriate channels.
   - Proactive service: respond to visible or latent customer needs based on (personal) data.
   - Customer engagement: involve the customer in the development of your channels, products, services and customer service (co-creation).

2. **Standardize and redesign customer processes**: Design services based on the needs and the "life events" of your customers. Organize the operation so that you are constantly aware of customer behavior and emotions. By setting up a professional customer feedback loop you can respond to both small and large customer signals as well as changes in the market. Stay in touch with the changing needs of the customer by regularly analyzing their journey. Do this together with the customer (co-creation), from the customer's perspective. Organize customer processes end-to-end with as few handover moments as possible. Ensure that each customer contact (incoming and outgoing) is recorded. If possible, deal with the customer request in the first contact (ensuring it is handled properly at this point) and if this is not possible then create a customer case. Monitor progress in the process, monitor actions, report if necessary and generate status information. Establish KPIs for customer processes and manage them actively.

3. **Develop competencies, attitude and behavior:** Ensure that everyone with customer contact develops customer- and solution-oriented skills. Be able to make real contact with customers, have the ability to empathize with the customer

based on a complete picture of the customer and their relevant life events. Ensure the adoption of new tools and discipline in recording customer contacts. Let your people play a guiding role in the digital world: where necessary, help the customer to become independent in the use of digital channels.

4. **Organize roles with a mandate:** Formulate and appoint the roles for customer process owner and case management. And organize data management for customer data and functional management of CRM solutions unambiguously in your organization.

5. **Develop information systems and technology**: Set up master data management (MDM): a single, leading registration of customer data. This will help to ensure that relationships are uniquely identified, the authentic source of all customer data is known, and the quality (accuracy, completeness, timeliness) is in order. Connect digitally to externally available sources where possible. Ensure that CRM functionality is always available everywhere and to everyone who has customer contact and that all channels are integrated with it. Preferably use the power of a state-of-the-art standard solution and take advantage of future developments. Build a 360° customer view, to be used by everyone with customer contact: your own employees, business partners and the customer himself.

None of these are easy tasks. Realization of a brilliant customer experience and a corresponding CRM implementation are among the most difficult organizational changes there are. It goes wrong quite often. If this is new to your organization, start with the basics: develop an initial version of a customer view, implement contact registration and case management, and integrate existing channels. Develop management information and improve current customer processes based on KPIs and customer feedback. If all this works, you can really start innovating: redesign customer processes based on life events and customer journeys. Start with those customer processes where you need to catch up or where you want to stand out in the market. After all, you can't do everything at once.

Example of a financial services company

A financial services company started an intensive program with the aim of making the change into a customer-oriented company. Given the importance, the CEO himself took ownership of this transition: "*We are doing this program from the belief that this will enable us to be more relevant to the customer. That's where our future lies, that's the business case*".

The transition affected the entire organization: redefining customer service, standardizing methods, focusing on customer KPIs, developing a more customer-oriented attitude and behavior of employees, introducing an integrated CRM solution in the cloud - everything was conceived, realized and implemented in combination. It started with the most important life events: moving to a new house, divorce, retirement and death. A total of approximately 1,700 front and back-office employees have customer contact, all of whom were included in the transition. Three million active clients are managed.

Approximately 50,000 cases per month are created and managed end-to-end. The effects are clearly visible after a while:

- Customers always get an answer or solution that fits their situation, because employees have the customer situation and the handling status available in the customer view. The effect of this is visible in a considerably higher net promoter score.
- Intermediaries notice improvements as employees are better informed and share more relevant information with them, such as leads and sales opportunities.
- Employees understand that, with so much customer information in one place, they have been given a powerful tool to help ensure that customers are satisfied.
- Managers can steer by customer cases rather than work backlogs. Greater transparency has increased the ability to learn.
- The organization achieves increased customer retention, more sales opportunities and cost reductions as a result of less rework, the phasing out of systems and supply chain optimization.

In addition, the "by-product" of such an intensive change program is not insignificant. More than 200 people worked together across business units in the program and various change competencies were professionalized.

7 Digital breakthrough 4: Agile and resilient organization

7.1 WHY IS AN AGILE AND RESILIENT ORGANIZATION SO IMPORTANT?

The previous chapters show that organizations are facing many changes and at a breakneck pace. In most changes, digitization and the deployment of new information technology play an important or even dominant role. There are quite a few demands on IT. By digitizing business processes, you want the lowest possible production costs in order to counter new market entrants. You want to offer excellent self-service capabilities to customers and business partners by means of web portals and apps. You want to use the large amount of available data to generate insights and set up AI processes. IT needs to accelerate the time-to-market of product development. IT must provide optimal support for your employees and for collaboration both internally and with business partners.

You also want IT to be delivered at a healthy price/performance ratio. It is important that information systems and networks are available, reliable and predictable, and that they comply with legal frameworks and the requirements of regulators. Your organization's IT function must play a prominent pioneering role in all of this. Your business strategy and ambition require a powerful IT function that can help realize them and ensure that you become fast, nimble and agile.

At the same time, we see that the IT industry itself is changing rapidly. IT is increasingly developing into a utility, a common infrastructure through which functionalities for the business, chain partners and end customers are developed. IT consists more and more of interchangeable components for hardware and networks, workplace automation, business applications (CRM, logistics, finance, human resources), data management and analysis, process control and portal functions.

Software development can nowadays take place without any programming (e.g., with low-code and no-code software), which can deliver an acceleration of 5 to

10 times compared to traditional system development. Many applications and IT facilities are available in the cloud, which can offer enormous advantages: easier upscaling or downscaling, automatic software updates, subscription-based billing so lower investments, better security, backup and recovery capabilities. And there are concepts and solutions available for digitizing the IT function itself (for example, IT4IT from The Open Group), so that IT delivery processes are automated.

The IT function within your organization must keep up with these developments or else it will miss the boat. If you have the ambition to move to phase 3, then your IT organization needs to be in line with this. Or rather, wants to anticipate it.

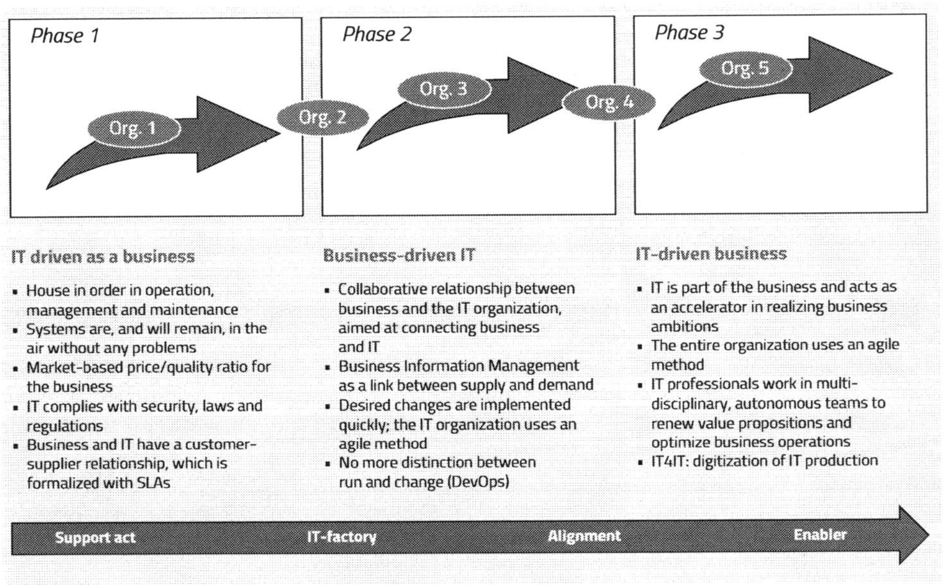

Figure 20 Stages of organization of IT

To achieve this, a paradigm shift is needed in how IT delivery is organized and managed in relation to the business. When we talk about IT delivery, we mean all IT-related processes, whether centralized, local or external. And the term IT is understood broadly, it is not just about technology but about the integrated approach of business and IT. After all, there are hardly any "pure" IT projects anymore. The majority of what are often called IT projects are in fact business projects with an IT component (heavy or not). The classic IT function must, therefore, change fundamentally, investments must be viewed more broadly than from the IT perspective alone and a different way of working together is needed:

Traditional	New
Building applications	Buying/defining functionality
Do it yourself	Directing execution
Exception is the rule	Standard is the rule
Designing and coding	Selecting, contracting and customizing
Investment proposals for IT deployment	Substantiated integral business case
IT costs as a black box	Transparency in costs and values
Acting as an IT expert	Interaction with business
Internal customer-supplier relationship	Together with the business focus on (external) customer
Self-evident right to exist External suppliers	Proving added value Collaboration/partnering

However, this is often not yet in place. There is still a lot going wrong with the current IT delivery processes and governance; the current IT function (and the demand role in the business) are not always sufficiently prepared for the future. A digital breakthrough is needed in order to become an agile and resilient organization.

7.2 PROBLEMS AND CHALLENGES IN IT PROCESSES AND CONTROLS

The business is disappointed on a regular basis about the missed opportunities and obstacles it experiences due to the poor or slow functioning of the IT function. Things tend to go wrong with the delivery of workstations, testing and moving into production, release management and renewal projects, which after huge investments end up delivering much too late or sometimes not at all. The price/performance ratio is not always in balance, and certainly in the eyes of some customers the IT department is perceived as too expensive for what it delivers. The business then sees IT primarily as a cost item that is difficult to influence. But when looking at things the other way around, the IT department is often frustrated about the vagueness and volatility of the business demands and frequently feels undervalued for what it has achieved through immense effort.

This situation occurs in many organizations. Common causes of these problems can be traced back to both the business side and the IT side:

- The demand-supply model, intended as a clear separation of roles between supply and demand, has often been implemented too instrumentally in daily practice. Instead of seamless collaboration, a waterfall has been created with too much distance between business and IT people.
- Departments where high speed is required no longer want to be dependent on their IT function and organize their IT outside of the formal paths ("shadow IT").

They hire IT specialists themselves, sign contracts with suppliers and buy standard solutions. They then ask the IT department, often to the horror of the CIO, to manage and integrate these solutions into the application landscape.
- Ownership and responsibilities, for example over applications and data, are not always clearly arranged. This results in discussions about costs, decision-making and responsibilities when problems arise.
- There is overdue maintenance, both functional and technical. Information systems operate with outdated versions in terms of technology and software. This doesn't only apply to existing systems, since even today systems are built using outdated technology, the legacy of the future.
- IT budgets are spent for the most part on improvement, management and maintenance; only a small part is available for renewal and innovation. At the same time the ambition of the business demands much more than that.
- The knowledge and competencies of the IT department do not sufficiently match what will be needed in the future. The organization is primarily geared to working with older techniques and systems and is not sufficiently focused on the deployment of new, modern technologies and innovative applications of IT.
- Execution power is low. Decision-making takes too long; everyone wants to have something to say about it. Projects do not focus immediately on executing the assignment but instead spend a lot of time initially figuring out things like standards, definitions, tooling and approaches.
- At crucial stages in strategic change, the organization is dependent on outsiders. Consultants and IT suppliers make themselves indispensable and are in control; there is insufficient direction from the organization.
- The culture that has come about is a consequence of objectives from the past: being in control, cost reduction and focusing on resources. People work from islands, think from the inside out and often reason from a technical point of view. Instead of strengthening each other and working together, too much energy is spent on convincing each other of opposing views.

The traditional separation between business and the IT function results in too slow a pace and a reactive, wait-and-see and internally focused organization. This often goes together with a lot of focus on the legacy environment and optimizing IT to be more cost effective. If your organization wants to move to phase 3, you must break through this vicious circle and the IT delivery process will have to be radically restructured and organized. Such that your organization becomes agile and can move quickly on the waves of all developments.

7.3 WHAT DO YOU NEED TO BECOME AN AGILE AND RESILIENT ORGANIZATION?

Reading tip:

Business/IT Fusion - Peter Hinssen, 2011

Today's technology giants, the well-known players from Silicon Valley but also the (often less familiar to us) parties from Asia, are among the most innovative companies in the world. The main distinction between them and existing corporates, the established ones, is that technology enables them to move faster, more flexibly and on a larger scale in the market. For these relatively new organizations, IT is not a difficult domain that gets in the way but instead an enabler, driver and inspiration for continuous innovation and adaptation to a changing environment. As a result, these players are not only successful in their "own" market segments, but they are now also breaking into segments that are new to them, developing digital business models for them and making life very difficult for the established order. The key question for existing corporates is therefore: how to operate and innovate as a tech company?

Example of a multinational in consumer electronics

There are examples of existing corporates that have succeeded in reinventing themselves in the area of data and digital. A multinational consumer electronics company with several product-related divisions had major problems with the way IT was managed. The business was complaining about the time it took to deliver new IT systems, the high costs and the technocratic and centralistic approach of IT staff. The business had all kinds of innovative ideas, but they always got stuck because the IT department smothered every initiative with endless discussions about compliance with standards, guidelines and approaches. There were many IT professionals, almost all working in one central IT department.

The management intervened and appointed a new CTO. The first thing he did was to outsource all non-core IT tasks, including most of the IT staff. Subsequently, the IT department was largely decentralized: IT people who spent most of their time working for one division were physically and managerially moved to this division. At the central level, a small and powerful management organization remained, with enterprise architects, security specialists and IT infrastructure specialists. Their core task was to facilitate the local IT staff as much as possible within jointly established guidelines. Guilds were set up with the aim of further developing knowledge and skills in the field horizontally, right across the company. In just a few years, the IT specialists have been transformed from reactive, pedantic and executive specialists into proactive and leading professionals in the business, who play a decisive role in change and renewal.

Successful tech companies, acting in phase 3, have organized their IT around a set of modular business services, executed by multidisciplinary value teams. Each business service consists of a logical cluster of products/services, processes, data, business know-how and associated technology. Each of these business services is managed separately and "assembled" to form the backbone of a company's digital capabilities. In the multidisciplinary value teams, business and IT people work closely together and have the decision-making power to act quickly. The value teams don't just work for the business, they are part of the business. There is no longer a separation between business and IT.

Value teams are concerned with introducing new propositions to the market, processing feedback from customers, business partners and internal users, and learning and improving in a continuous process. If your organization's primary processes are largely digitized, then this has become the core of your organization.

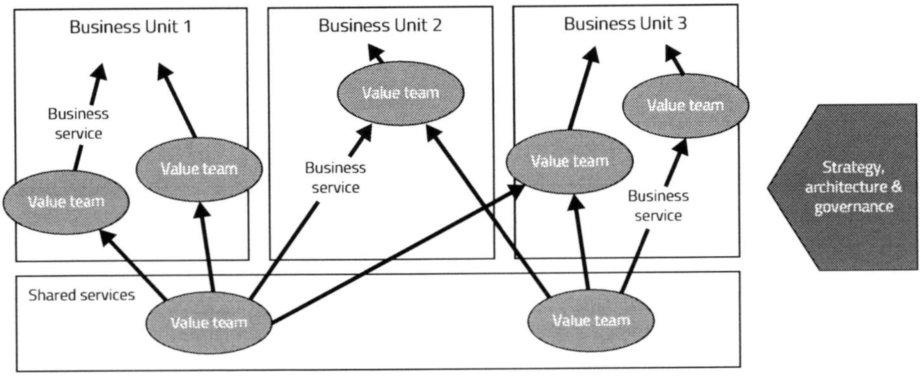

Figure 21 Value teams deliver business services

Value teams focus on developing business solutions to serve customers (internal or external) and to deliver them to other value teams. In the latter case, the business service is organized as a shared service, for example, common input and output processes or a shared data ecosystem. Value teams operate as independent entities that bring together products/services, processes, data and technology and are empowered to act quickly. The teams are responsible for the entire IT chain, from idea to realization, from raw material to final product. They are run by a product owner from the business, who has budget, sets priorities, formulates high-level requirements and takes ultimate responsibility for the business service. Value teams are cross-functional, with business representatives, IT people and functional specialists (such as marketing, data analytics, legal etc.). They work in an agile manner, deliver the solution together, enable continuous innovation and develop and manage all necessary IT.

This modular, service-based IT setup adopted by tech companies is what allows them to accelerate and innovate. They can experiment, fail, learn and scale quickly, and they can bring their products to market many times faster.

Example of a banking organization[7]

A large bank operating globally created around 30 business services. One example of such a business service was "payments", which consisted of more than 20 separate applications that had previously been managed independently of each other. It was decided to merge all IT people working on the development and maintenance of payments with the corresponding people on the business side.

Under joint business/IT leadership, this entity was given the mandate to prioritize, modernize and allocate resources. The team began to execute the payments business service as an internal business that served all customers within the bank. The team revamped the entire IT landscape in a short period of time. This business service-based way of working was then successfully rolled out across the group within a year.

The approach is in stark contrast to the way work was done before; setting up an organization to revamp IT or execute a major IT program was usually a multi-year undertaking.

The value teams have maximum autonomy within clear frameworks. For the latter, you need a small, powerful function that steers horizontally: *strategy, architecture and governance*. It develops the digital strategy and sets up and applies the enterprise architecture. And it develops the standards for security and privacy, IT control (monitoring the effectiveness of IT delivery) and the sourcing strategy. It is not a police officer, but is fully focused on clarifying the direction, architectures and policies ("guardrails") and removing obstacles, so that the value teams can fully focus on achieving their mission.

SAFe: Scaled Agile Framework[8]

Many organizations have moved to an agile way of working in recent years. This often started on a small scale within the IT department, with the aim of delivering customer value in an accelerated and flexible manner via a small team. However, scaling up requires more and more coordination between teams that depend on each other.

SAFe has been developed specifically for this purpose: it is a collection of methods, principles and approaches for the problems that ossified organizations experience in their desire to become more agile.

SAFe is a framework designed for organizations where large groups of employees work together on complex products and services in changing environments. This requires a structure that dynamically handles opportunities, problems and possibilities. SAFe incorporates concepts from Scrum, Lean, Kanban, DevOps, Lean Startup and Design Thinking. This integrated overall vision, with ideas on project management, product development and software development, has led to SAFe now being the leading framework for large organizations. SAFe has three levels:

---

7 Based on: Oliver Bossert and Driek Desmet, "The platform play: how to operate like a tech company", February 2019, www.mckinsey.com

8 See also: www.scaledagileframework.com

1. Portfolio: this refers to the translation of the business strategy into the digital strategy and is where the strategic themes are defined.
2. Program: at this level, the digital strategy is translated into priorities for the coming period and dependencies are derived.
3. Team: at the team level, work is done in short periods of time (sprints) to produce results that add value immediately.

The key messages of successful agile implementations are: "fail fast, respond to change and deliver faster and more often". And: "perfection is a fallacy that leads to too much planning, procrastination and cancellation". Agile is about focusing on delivering the best that is possible in a given period of time.

7.4 HOW TO BECOME AGILE AND RESILIENT?

Transforming the organization towards the outlined model is no easy task. Think big, act small. Start there, where speed must be the highest priority and where there is pulling power in the organization. This is often the case in Marketing and Sales. Introduce agile working, start with self-organizing and multidisciplinary value teams, responsible for the result and thus for the planning and distribution of work and progress. Appoint product owners, invest in ownership for processes and information systems that belong here, transfer IT staff and IT budgets. Work with fixed budgets, within which the teams set their own priorities. Have the teams start by working out a concrete digital strategy for their domain. Create the first successes here, learn and improve, show that it works. Make sure that the following things are put in order step-by-step:

- Work according to the principle of "see-feel-change" or learning by doing. Define short sprints (iterations), stay on the ball. Set up living labs, prototypes and pilots to see if you are on the right track. Take the next step based on practical experience.
- Put the mandate where the expertise is. With those who are in the best position and have the appropriate knowledge to make the right choices. These are not always the generalists (the managers) but often the specialists or employees who are closest to the customer. Nurture the expertise of these specialists. *"It doesn't make sense to hire smart people and tell them what to do; we hire smart people so they can tell us what to do"* (Steve Jobs).
- Minimize dependencies and coordination. Make steering committees superfluous where possible; organize accountability in the line organization as much as possible. Arrange for those involved in value teams to be given authority so that they can move quickly.

- Establish a horizontal governance: strategy, architecture and governance, and develop the overall vision, digital strategy and architectures (guardrails) here.
- Develop cross-organizational employees in terms of knowledge and skills: new ways of working and technologies, flexibility, sense of responsibility, collaboration, creativity and entrepreneurship.
- Provide a leadership style that suits the new way of working. Managers who give space and trust to employees, are facilitators and connectors, lead by example and give sincere compliments.

The transformation to an agile organization is far-reaching and causes a major organizational and cultural turnaround. Where IT used to be the support for the primary process, in the new world it is deployed for the smart digitization of processes, creating a data-driven organization, realizing a brilliant customer experience and renewing business models. If this is not happening fast enough, it can help to (temporarily) appoint a Chief Digital Officer with one mission: break down silos and implement a fully integrated digital strategy.

8 Digital breakthrough 5: Digital business models

8.1 WHY ARE DIGITAL BUSINESS MODELS SO IMPORTANT?

There you are, with friends in an Uber cab being taken to the Airbnb for a short vacation in Portugal. During the trip you listen to your favorite artist on Spotify with your friends. You hesitate, should you go to a concert after all?

More and more we see digital platforms appearing around us, without us being aware of them. You can see this in the example above: on Spotify's platform, where artists and music lovers meet. On Uber where drivers and travelers meet. And on Airbnb where homeowners and tourists meet. These are three commonly cited examples of digital platforms that have proven to be fast and successful. This chapter will take you through the fifth digital breakthrough: the changing world of digital business models. Several types of digital business models can be recognized; however we will now mainly focus on that of digital platforms, one of the most important and revolutionary business models of recent years.

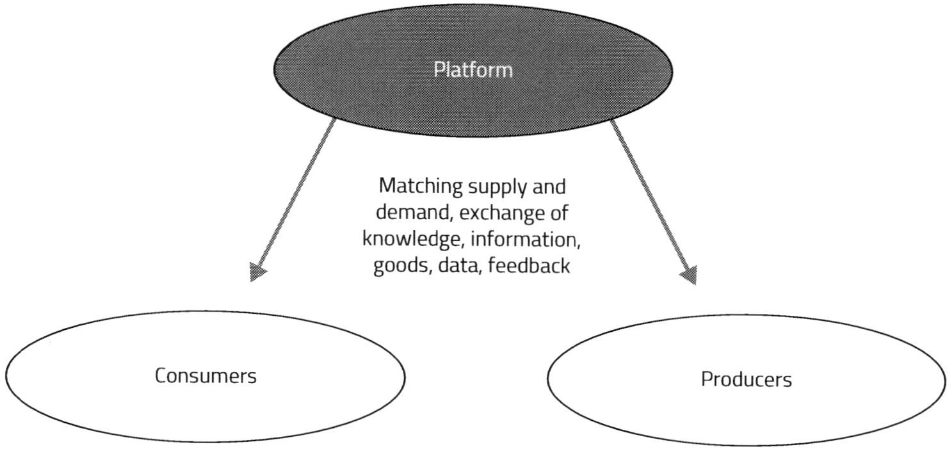

Figure 22 Essence of a digital platform

The essence of a digital platform is that of a digital marketplace, where supply and demand are brought together. It is not a technology but a business model, with a new form of value creation, innovation and revenue model, where technology and data play an important role as enablers.

> **Reading tip:**
>
> *Platform revolution*; Geoffrey G. Parker Marshall W. Van Alstyne 2017

Marketplaces have been around for much longer, the difference with digital platforms is that we now have digital technology. We can communicate with each other all over the planet and we can complete transactions almost instantly. Platforms are often disruptive in nature, having a major impact on traditional organizations, the established ones. Successful digital platforms grow not linearly but exponentially. Where traditional companies mainly focus on revenue and profit, platforms focus on attracting more producers and consumers. Especially at the start, growth is the main driver of a platform.

Digital platforms have some major advantages over traditional companies:

- Platforms use resources from outside and, therefore, incur relatively low costs compared to the established order. The costs of developing a platform can, of course, be substantial. However, once the platform is in place, it is mainly about further development; the continuous search for (extra) value for consumers and producers.
- Platforms can scale more easily. Because a platform focuses on a scalable and sustainable IT architecture and solution from the start, it can grow faster and does not create a slowing legacy as we often see in traditional organizations. Moreover, national borders are fading; the whole world is accessible.
- Platforms can better respond to customer needs through data. A digital platform generates a lot of data and will use this data for analyses and predictions. These insights are used to respond to the needs and wants of consumers and producers, which is crucial to continuously adding value for users.

How should we position digital platforms compared to traditional business models? And should you also get into the "platform business" or should you join it as a producer?

The model in Figure 23 outlines the total playing field of what you can focus on as a company, with vertically the value chain (distribution, customer service, product development, production and purchasing) and horizontally the customer journey (orienting, searching and comparing, buying, using and evaluating). Of course, this is a simplification of the much more complex reality, but it helps in thinking about digital strategies.

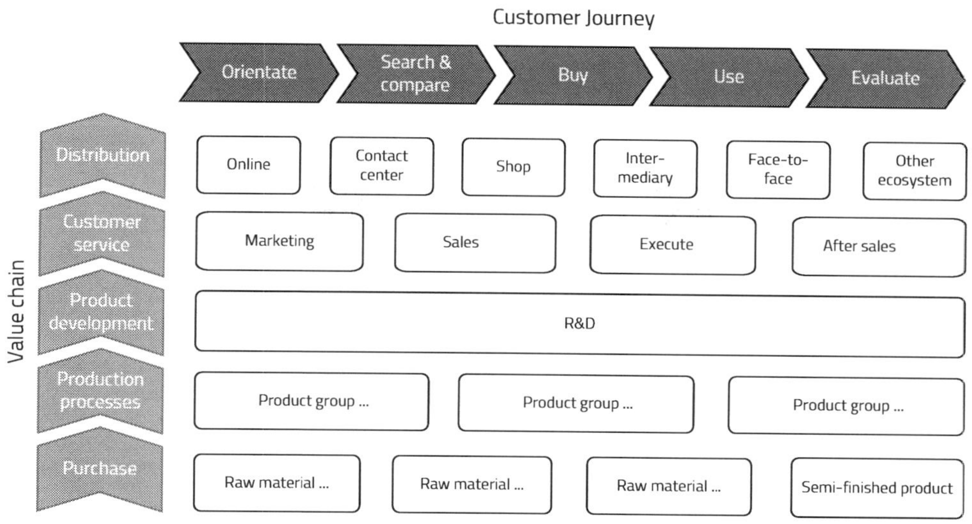

Figure 23 The total playing field; simplified model of reality

Many organizations cover a significant part of this model, see as an example "spot" A in Figure 24. They invent, develop and produce products and services. They serve the customer, through various distribution channels, often from beginning (customer need) to end (providing aftercare). But suppose a producer enters the market with the sole aim of producing one of your product groups as cheaply as possible (circle B in Figure 24)? With fully digital processes, huge purchasing discounts due to large volumes and a scalable architecture. Can you then compete with that?

And suppose a digital platform enters the market, bringing together supply and demand for your type of products and services (circle C in Figure 24)? With distribution costs that are a fraction of your contact center, stores or intermediary network? Who "steals" your customer contacts including all valuable data. Is it strange that many retailers in fashion and electronics, travel organizations and media companies have been wiped out of the market?

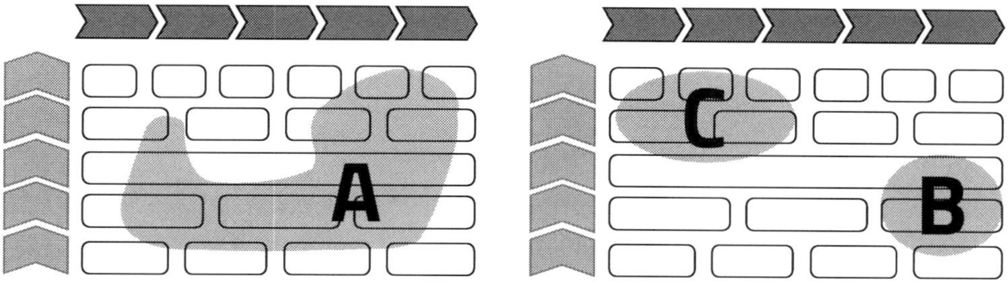

Figure 24 Example positioning of organizations

There has been an unbundling of the value chain for quite some time. Organizations will have to quickly choose which business model they want to focus on; wanting to do everything is no longer an option. It is certainly possible to opt for multiple business models if they are organized and managed separately. After all, each business model has its own revenue model, different KPIs to steer by and different requirements in terms of organization, working methods, leadership, competencies and IT structure. However, if you mix different business models, it will eventually go wrong.

Example of digital business models at a financial institution
A financial service provider has insurance, mortgages, pensions and banking services in its product range. All products and services are developed and administered in-house, and distribution takes place via intermediaries and the direct channel, using a contact center and the internet. The market is difficult: much competition, low margins and new entrants. Legacy problems are blocking progress.

Based on a strategic repositioning, it was decided to set up a financial platform: a single (online) point of entry where customers can arrange all their financial affairs, where supply and demand in relation to finance are brought together, and which helps customers to make the right financial choices. This covers the full range of payments, savings, investments, insurance and mortgages. Using the latest technology, based on intelligent algorithms, incorporated in a user-friendly site and app with handy tools. A platform that continuously matches the best offers from financial providers with the personal situation of the customer, completely transparently and independently. Focused on customer needs: "Do I have my financial affairs in order? And my insurance policies? And the mortgage? Am I saving smartly enough? And what is the best thing I can do for a good pension? Could it all be simpler and more accessible?"

At the same time, it was decided to dispose of the administration of "commodity" product groups such as various non-life insurance policies, in which the company does not excel enough, where there is too much competition and the margins are too low. The focus will be on those product groups where knowledge and experience make the difference: pensions, income and financial advice.

The financial platform and the various product groups are positioned, organized and managed separately in the group.

It is expected that before 2030, a good portion of value creation by companies will be in the form of a platform model. New platform companies are being created every day. Many do not survive, others do. This process continues with no signs of slowing down, making it highly likely that further changes will be imminent. Organizations currently operating in stage 1 or stage 2 would do well to orient themselves on how they want to fit into the platform economy. Such an orientation may result in: becoming (partly) a platform business yourself and/or becoming a producer in a platform business model of others. Or do you think, by defending and building on your current position in various ways, that a platform has no chance of succeeding in your market? For example, if there are extensive regulations, high entry costs into your market and little effect of digitalization. Who knows?

8.2 PROBLEMS AND CHALLENGES WITH DIGITAL BUSINESS MODELS

If a corporate from the established order wants to set up a digital platform, there are a lot of advantages to be gained from organizing this completely outside of the existing business. New branding, clear positioning in the market, organizing as a Lean Startup, attracting top talent. And developing the platform directly under an architecture, without the ballast of legacy. However, you should also bear in mind: you start without existing customers, you must set everything up from scratch and you will have to arrange the financing of this properly.

The alternative is to organize a digital platform "close to" or even within the existing business and to use the valuable assets that are already there. You can use (the data of) your existing customer base; you have potential visitors and traffic on your platform. If the current brand name already has a positive customer experience, you can hitchhike on this and build awareness faster. You can allocate top talent from within the existing organization. But this variant will also have its drawbacks: there will be pressure to fit in with the governance of the corporate, which can be fatal to the required speed and clout. There will probably be a focus on profit contribution rather than on growth sooner than desired. And despite all good intentions, you will undoubtedly have to deal with accumulated legacy, before you know it all the attention will be on solving problems of a phase 1 and phase 2 organization.

There is no one-size-fits-all solution to this issue; for each situation, the benefits, drawbacks and risks will need to be identified. And mitigating measures must be taken to contain the risks. Even if you have gone down this road and have started setting up a digital platform, you will still encounter many issues:

- The chicken-and-egg problem: how do you attract parties on both sides to your platform if nobody is using it yet? You'll have to do something, especially at the start, to attract users: for example, by (temporarily) providing extra value, rewarding users for participating, offering incentives or creating engagement.
- How do you achieve growth on the digital platform? Growth is realized through "same-side effects" (producers attract other producers and consumers attract other consumers) and "cross-side effects" (new consumers attract new producers and vice versa). This ultimately creates a snowball effect; once the ball starts rolling, a platform will grow substantially. But getting this flywheel going is quite a challenging process in practice.

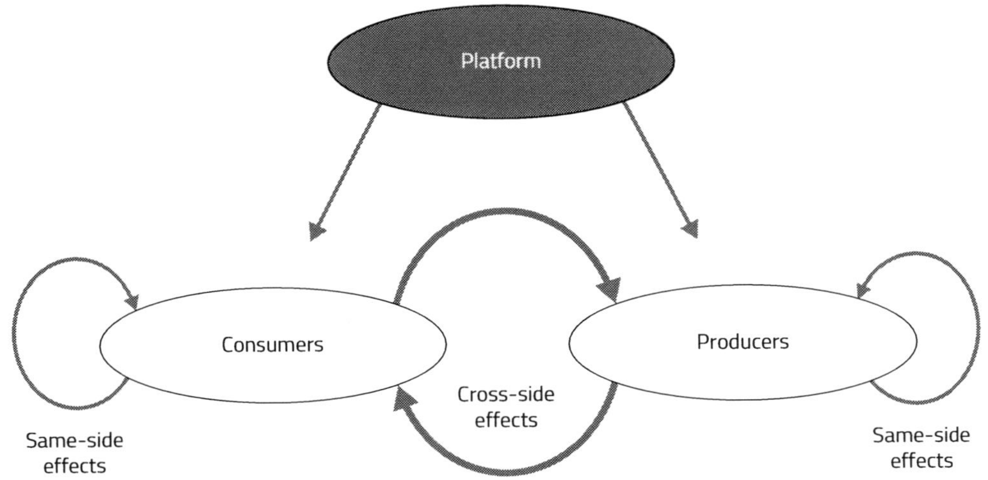

Figure 25 Growth of a digital platform

- Whilst growth is the primary driver of a digital platform, ultimately it must also make money. Many digital platforms struggle with the revenue model. Possibilities include the freemium revenue model; here users get free access to basic functionalities but must pay to access more advanced features. Other options include revenue from online advertisements, a commission on the sale of a product between producer and provider, or the sale and provision of (anonymized) data to third parties. These kinds of choices are very important; if they don't like it, users will have left your platform before you know it.
- Regulation and ethics: many digital platforms have an open nature, users can add value, for example content and functionality. Think Wikipedia or various social media platforms, or an open software platform. But this can also lead to wrong or even malicious things being done with them. There must be effective rules, security provisions and quality control to manage this.
- In addition to consumers and producers, there is a third group that is indispensable when developing a platform: professionals. You must be able to attract scarce online marketers, CX specialists, content managers, data analysts, IT architects and platform developers. And thus compete with startups.

Platforms in the public domain?

The phenomenon of digital platforms is not something that only commercial enterprises have to deal with. The Dutch government is seen as a front runner when it comes to digital services compared to other European countries. The creation of government services will increasingly be a matter of co-creation with other players: companies, social organizations and citizens. Here, commercial supply and demand are not linked together but the various players make use of each other's resources, time, knowledge and possibilities. This has a major impact on how the government currently works and is organized. The government must take a leading role in this and not only facilitate but also initiate the cooperation of various parties. In this way it can fulfill an important social task.

Examples of government platforms include:

- The Cultural Heritage Agency of the Netherlands promotes cooperation in the field of cultural heritage, archaeology and art in the Netherlands. By organizing this information on a platform, supply and demand are brought together. The RCE has initiated and organized these collaborations so that many types of users can now use this information effectively.
- An example of a governmental cooperation platform is "Square Government". Here the government offers the possibility to meet digitally and to make use of discussion forums and blogs. Many different government agencies use this platform to exchange professionals, share knowledge safely and develop projects.
- Another example is the UWV "WORK Company". The UWV (the Dutch Social Security) has an important role in linking labor supply and demand. By establishing a platform, employers and job seekers can more easily meet each other. This takes place without the input and intervention of the UWV and without any commercial interest. On this online platform, where user-friendliness is paramount, citizens can therefore look for a job faster and more efficiently. In addition, the UWV offers more support services via this platform, such as writing a CV and finding information about the job market.

8.3 WHAT DO YOU NEED TO SET UP A DIGITAL BUSINESS MODEL?

The development of a digital platform starts with a clear strategy and scenario analysis. Why do you want this, what are the goals you want to achieve, what are the strategic choices and the starting points? How innovative is it, are there competitors, either domestically or internationally? However difficult, there will have to be a positive business case, otherwise there is little point in investing money and energy in this. The design of a platform can then be approached from various angles:

Design distinctive customer and supplier propositions

You will need to have a good picture of the target groups on both the consumer and producer sides. What are their needs, how are you going to fulfill them with the platform, what are the customer promises? The first contours of the propositions must be clear, so that you can quickly launch them into the market as a "minimal viable product" (MVP). What are its unique selling points (USPs), why should users go to your platform? What pricing model are you going to use?

Example of an online bank
An online bank in Germany has formulated six USPs for the consumer part of its platform:

1) Making it easy: fun and easy for everyone to use. Speaks the language of the customer. Is sympathetic and approachable. Be able to take care of everything immediately, eliminate worry.
2) Offer benefits: provides insight into what benefits the customer can obtain. With the possibility to arrange it directly. A smart subscription provides access to a wide range of services, including digital help.
3) Personalization: we know the customer. We therefore also provide specific benefits derived from personal insight and pricing.
4) Being proactive, we approach the customer: "you can achieve a higher investment result", "now is the best time to refinance your mortgage", "you will probably have 75 euros left this month to put aside".
5) Digital with a human touch: with increased digitalization, human contact becomes increasingly valuable. Computers help with personalization; employees make it personal.
6) Be flexible: the platform can be expanded with (external) applications that help customers budget, save money, save smartly, give insight into their spending patterns, organize their financial records, etc.

Establish an agile platform organization
You will have to make choices about how and where to position the digital platform in your organization. As a separate legal entity or "close to" or even as part of the existing organization? Do you link up with existing staff and support functions of the corporate, or will you organize these yourself? Whatever choices you make, the management and working methods will have to be fully in line with those of the "agile organization" described in Chapter 7. And the processes must be maximally digital: transactions are processed with "straight through processing". The platform organization is not concerned with executing processes but with continuously creating value for consumers and producers.

Develop people and competencies
Characteristics of the corporate culture will be: empathize with platform users, a learning organization, entrepreneurial and decisive, creative, focused on collaboration. Platform leaders are visionary, give direction, facilitate, connect and challenge. You are attractive to the top talent in the market: online marketers, UX specialists, data analysts, IT architects and platform developers. In other words, for professionals who can also be found at startups.

Make use of the data you generate

A platform generates masses of data about the users of the platform. With the help of this data, you get to know these users and it becomes possible to serve them personally. You can make use of all the data in combination with smart algorithms. Data and data analysis are therefore at the heart of the digital platform; the IT architecture must be data centric.

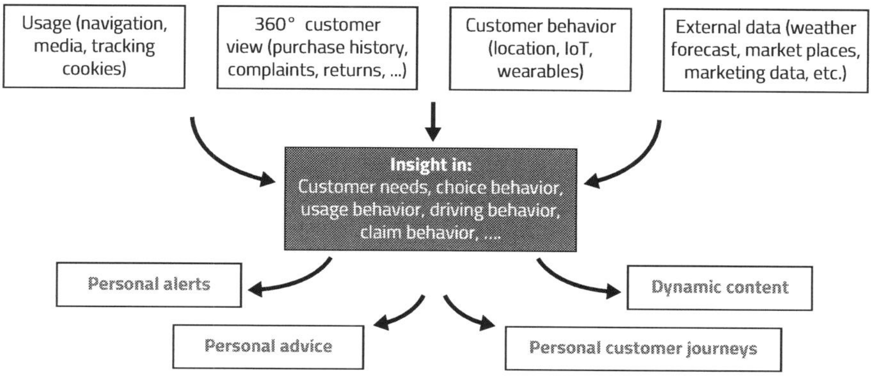

Figure 26 Use of data in a consumer platform (example)

Realize an open IT architecture

A digital platform consists of several IT components, which optimally support the customer journey. There are connections with consumers through all kinds of channels: directly, with their own website or app (B2C or B2B), linked to a partner (B2B2C) but also for example C2B2C, think of consumers who make music for each other and can be discovered by record companies. A platform can also be linked to another platform. The IT architecture of a platform is flexible and scalable and can grow with dynamic and changing environments. Components in the IT architecture are decoupled using standardized APIs.

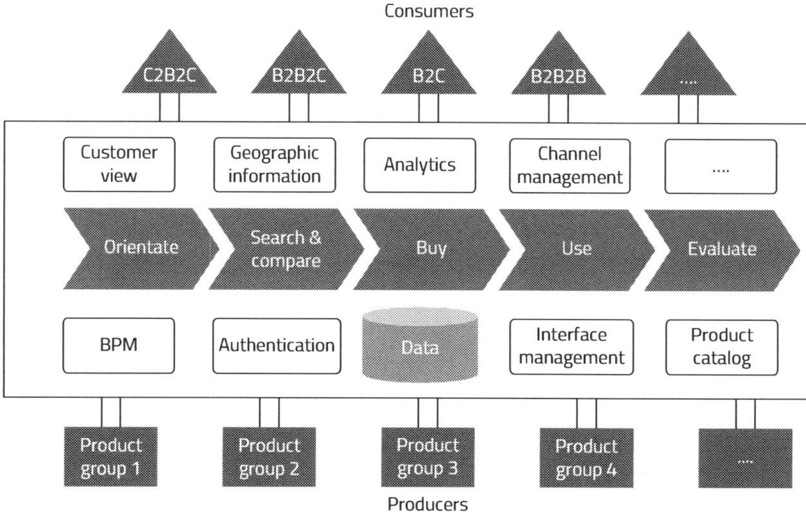

Figure 27 Logical platform architecture

8.4 HOW TO DEVELOP A DIGITAL BUSINESS MODEL?

It should be clear by now that if you want to build a successful digital business model, you will have to be reasonably advanced in the previously described digital breakthroughs. You have data management and data analysis in order as a capability. Your organization is mature in digitizing processes and you can deliver an excellent customer experience. If your organization is still in phase 1, do not start working on digital business models yet. But if you already have the characteristics of phase 2, then you are potentially ready.

You can also choose to take on the role of producer in the world of digital platforms. In that case, the key is to standardize your products and services to a large extent, simplify them, set them up modularly, create scale and thus a low cost.

The same applies when developing a platform as a digital business model: start small, test in the market, incorporate user feedback and scale up if it catches on. In doing so, follow the process shown in Figure 28, not once but continuously.

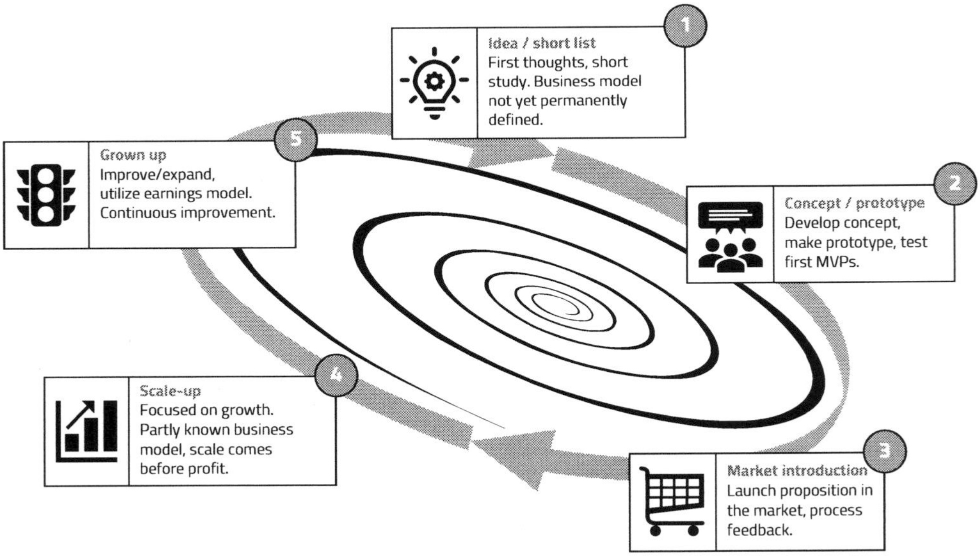

Figure 28 Digital platform development process

Select and prioritize the development of the platform by continually reviewing goals and criteria such as:

- Growth: quickly attracting many users and generating traffic.
- Distinctive: the ability to be truly innovative, different from others, with the functionality offered.
- Virality: the ability to leverage the networks of ambassadors and influencers to gain new customers.

- Customer experience: ensure it contributes to a positive customer engagement and high NPS.
- Short go-to-market: consider manufacturability, speed from idea to market launch.
- Revenue model: can you make money with it in a period of say three to five years?

The ranking and weighting of these goals and criteria is not fixed but will be determined per situation. So, no blueprint and roadmap in advance, here too the rule is: decide and adjust in a short cycle.

Lean startup approach is also useful for corporates[9]

Reading tip:

The Lean Startup; Eric Ries, 2017

A Lean startup is a startup company with an innovative idea, which experiments with a small team, gathers feedback from the market and improves its proposition based on experience. The product or service is both scalable and repeatable, enabled by the latest technology. The future of a startup is uncertain because it is not known in advance what customers will think of it. In fact, a startup often does not know who its customers are, and therefore whether the right customers are being approached.

Whereas the traditional waterfall approach delivers a complete product in one go, and according to agile working a total product is developed in small steps, a Lean startup uses a new form of product development. Firstly, the minimal requirements are determined, and these are tested in the market as quickly as possible. What works, what doesn't, what needs to be changed; that's what you want to know as soon as possible. You find out by introducing a prototype of the product, the very smallest part of a product that is usable for the potential customer: the minimum viable product (MVP). Based on this "mini product" you can find out how customers react. Startup founders therefore prefer to put their time and energy into developing the product, rather than writing a comprehensive business plan.

By measuring and incorporating feedback from the market, you can improve your product or service step-by-step. For every phase of the growth of a Lean startup, there is something that can be improved. The trick is to measure in such a way that you see the result of changes in the short term. Only when the product catches on with the target group does it make sense to put money into (large-scale) marketing, sales and production. By then, the product will often look different from what was originally conceived. So, it is more important to just make a start and leave the "perfect plan" for what it is. The approach of a Lean startup is excellent for both startups and existing companies.

9 Retrieved from: www.sixsigma.nl

9 Moving successfully from digital strategy to execution

9.1 DIGITAL STRATEGY AS PART OF THE CHANGE CYCLE

> **Reading tip:**
>
> *The Execution Premium*
> Robert S. Kaplan, David Norton, 2008

We started in this book by looking at the drivers of change. We have seen that digital technology plays a prominent role in many of these changes. All this has been put into perspective by the introduction of the phase model for digital transformation. Five dominant themes have been recognized, in which digital breakthroughs are needed: data-driven organization, smart digital processes, brilliant customer experience, agile and resilient organization and new digital business models. By developing a sophisticated digital strategy, you can tune and elaborate these themes for your organization and complement them with other relevant topics. All themes are linked to your business strategy, maybe your business strategy has even been changed significantly as a result.

But the hardest part is yet to come, namely the successful execution of the digital strategy: the "strategy-to-execution (S2E)". In other words, getting all the ambitions and intentions in place and harvesting the successes.

So, the digital strategy is just the beginning. You will need to have the entire change cycle in your organization well in place. Focus on the changes that create the most value. Organize executive power to implement changes quickly but also with the right quality, under architecture. And your organization will need to have a large learning capacity to make short-cycle adjustments, to learn from both failures and successes.

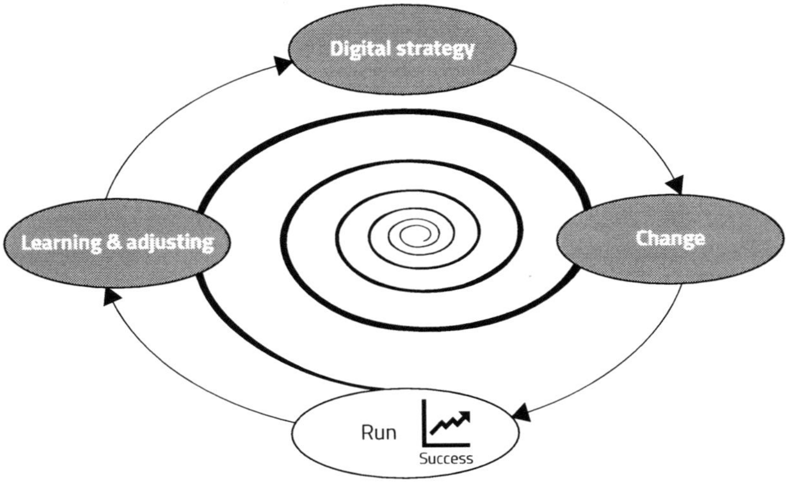

Figure 29 The change cycle of digital transformation

The change cycle of digital transformation has four steps:

1. The **digital strategy** is the starting point of the cycle. The digital strategy formulates the ambition, goals, future scenarios and digital themes. Considerations are made about how and where to deploy digital technology in the development of new business models, innovating propositions, improving customer relations, new distribution concepts and digitizing business processes. The result is a feasible and achievable digital roadmap.

2. In the **"change"** step, the digital strategy is brought to fruition. Much is already known but not everything is in focus. We don't always know the impact of the choices we make. Business models have not yet been firmly defined, we don't know whether the ideas will catch on in the market, or whether the ambitions and goals are achievable.

Innovation funnel
Innovations are best managed using an innovation funnel. The funnel starts with a longlist containing all kinds of possible digital initiatives, both ripe and green. Based on pre-agreed criteria, initiatives are evaluated and either taken a step further or stopped. Examples of criteria are: fits the vision and mission, there is significant market demand, it involves a potentially distinctive proposition, the sponsor and initiator are present and committed, etc. In the funnel there are formal "checkpoints" where an initiative is tested against the criteria. The result is: proceed to the next phase or stop. The criteria help to move the initiatives in the funnel from idea to concept to prototype or "minimal viable product (MVP)" and then to market launch.

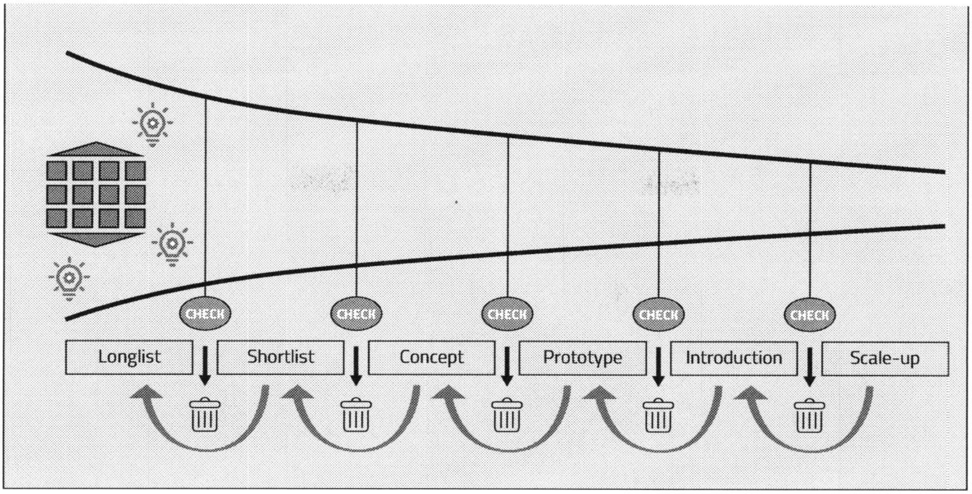

Most of the formulated changes in the digital roadmap will be sufficiently clear to get started. Renewal and changes in propositions, market approach, business processes, organization and digital technology will all be developed, tested and introduced in a coherent manner. New propositions and customer processes will be marketed and adjusted based on feedback from customers, employees and other stakeholders.

3. In the "**run**" step, after sufficient proven success, the changes will be structurally integrated into the organization. The changes will be scaled up and the business model will be further improved and expanded. This should be a continuous improvement process, an ongoing effort to adapt products, services and processes to the requirements and wishes of the environment.

4. Finally, **learning and adjusting** takes place in the cycle: a coherent evaluation of all digital change activities and of the performance in the achieved situation. In other words, this is where the effectiveness of the digital strategy is measured, and any adjustments are made where necessary. The effects on previously developed business cases are determined: are the predicted benefits being realized ("benefits tracking")? Based on the findings, the digital strategy is reassessed and adjusted. The circle is complete.

The prerequisite for successful digital transformation is that all the above steps are balanced, simultaneous and professionally embedded in the organization. This means that the steps are not linear but rather simultaneous, continuous and iterative. Not an annual exercise but a cycle that is woven into the organization as a continuous process. The digital strategy must be constantly adjusted, because organizations and their environments are constantly changing. Digital transformation is, therefore, not something one-off but rather a continuous process. Nevertheless, if you don't already have a clear digital strategy, it's a good idea to start putting it together in order to get the flywheel turning.

9.2 PRIORITIZING: AN INITIAL RANKING

Using the framework for business transformation, all relevant digital themes including associated change initiatives are mapped out and, if desired, supplemented by other IT-related change activities that are at play in the organization. In the digital roadmap, choices are made in relation to what you select in order to go from today to tomorrow. All changes are included that are necessary to realize the digital ambition.

You don't eat an elephant in one sitting, so it's wise to divide the roadmap into small manageable pieces and continually update it based on successes achieved and new developments. The roadmap is aligned with the business goals and focused on changes that deliver the most value. It is important to prioritize and make sense of dependencies. The value/complexity matrix, shown in Figure 30, helps with this prioritization.

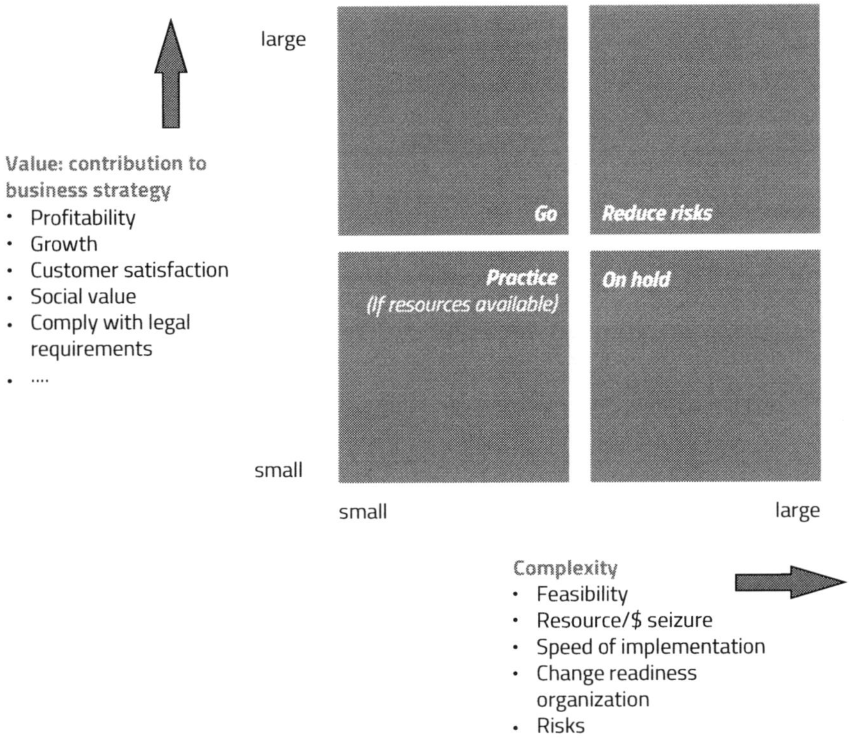

Figure 30 Value/complexity matrix

To rank all digital initiatives, they are weighted using two dimensions:

1. **The relative value,** or importance to the organization: the contribution to the mission, goals and strategy, such as return, growth, customer satisfaction, social value, meeting legal requirements, etc. In fact, this is a first rough business case for each digital initiative.
2. **Relative complexity**, which looks at feasibility and scope (in terms of clarity of task, expected capital expenditure, political influences, available knowledge/ capacity, use of new technologies, maturity of organization, and number of different interest groups).

Based on the "scores" of the change initiatives on these two dimensions, the priorities can be set:

- Relatively high value, low complexity: implement immediately; no discussion.
- Relatively high value, high complexity: reduce risk first, for example by splitting up the initiative and re-scoring the parts on the dimensions. Or by taking an incremental approach to the initiative and experimenting.
- Relatively low value, high complexity: don't do it, discard.
- Relatively low value, low complexity (mostly "low hanging fruit"): implement if resources, time and capacity are available.

It may help to categorize all digital initiatives into three "buckets".

1) Maintain, belonging to the strategy "optimize existing business for as long as possible and desirable".
2) Growth, belonging to the strategy "create more relevance for customers and ensure growth in existing business".
3) Innovation, belonging to the strategy "develop new business models in which you connect customer needs and solutions".

You could agree in advance that the total change resources (budget, knowledge, capacity) will be divided among the three "buckets" according to a distribution key, for example 30% to maintenance, 50% to growth and 20% to innovation. Or using a ratio that best suits your organization. In this way you are "forced" to create room for innovation and growth, and you avoid a situation where all resources are spent on maintaining and optimizing your existing business.

Of course, none of this is overly complex and there are gray areas to be identified. But rationalizing the discussion helps to focus on the goals, value and scope of the digital initiatives and assists in arriving at the right choices during the process.

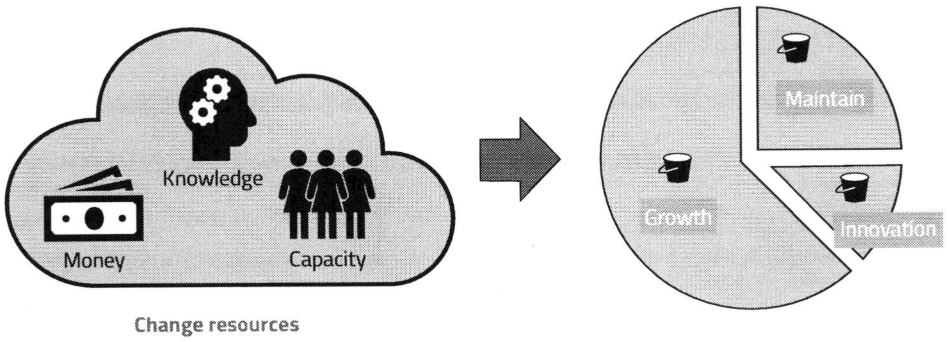

Figure 31 Distribution of change resources across buckets

9.3 EIGHT CORE PRINCIPLES FOR SUCCESSFUL CHANGE

The digital strategy has been drawn up. Priorities are set, the digital roadmap is clear, the organization is ready. In bringing the strategy to fruition, there are several factors that can influence success or failure. Many have already been covered in previous chapters; below is a brief summary of the most important success and failure factors:

1. **Digital transformation = business change**
   Too often in digital transformation, one looks to the IT department with the result that changes are approached from a technology perspective and not enough from a business perspective. This does not go well for long. IT projects don't exist at all, most of what are typically called IT projects are in fact business projects with an IT component (heavy or not). It is not about building and implementing IT systems but about new business models, renewing and improving customer service, optimizing processes and supporting employees, business partners and customers with effective tools. Working methods will change, people need to be trained and coached for this. These kinds of projects need to be approached in a multidisciplinary way and managed from a business perspective.

2. **C-level ownership**
   Digital technology affects the entire organization; everyone is involved. In order to break down existing complexity (and prevent new complexity), hard choices sometimes have to be made. This requires broad ownership at the C-level and the ability to escalate and adjust quickly if necessary. It is essential that top management is actively involved, propagates the importance and immerses itself in the matter. The most important tasks are: spreading the vision, connecting, giving front runners the podium and removing any obstacles to speed.

3. **Change under architecture**
   We have previously discussed how corporates, established organizations, are often burdened by a huge legacy problem. The enormous complexity of products, processes and the IT landscape threatens to block the realization of the business strategy. The way to grow from the historical situation to the new world is to "change under architecture". Use architecture as the basis, the foundation for change. In the physical world of building, "rebuilt under architecture" means that the rebuilding was designed and supervised by a qualified architect. This is also necessary in digital transformation; only then will you achieve the sustainability and flexibility that gives you an edge over the competition.

4. **Work incrementally on business value as well as digital capabilities**
   Start in a practical and small-scale manner: start quickly, learn by doing and in this way achieve tangible results quickly, in small steps. Evaluate and improve continuously. If successful, implement the results more widely in the organization. Seize the momentum to build and professionalize the core competencies of employees. In this way, the digital capabilities of the organization will grow.

5. **A good start is half the battle**
   Many digital transformations fail at the outset. This happens regularly because a start is made without it being clear what must be delivered. Halfway through the process, the transformation runs into problems because of this lack of clarity, resulting in higher costs and a considerable overrun. Before the start of any initiative, the overall objective must be clear to everyone: this includes the goals, the results, the scope, a first rough business case, and the organizational, technical and financial consequences. No extensive studies, just a short, sharp definition of the goals and mission.

6. **Holistic thinking, autonomous action**
   You create speed by giving digital initiatives space, freedom and autonomy. But digital changes are not isolated from each other. There are dependencies in terms of content, whilst we often need to call on the same experts and the total change budget is limited. It is, therefore, a good idea to direct and monitor the realization of digital ambitions with the help of change portfolio management. This can be done, for example, by developing a change calendar with milestones and dependencies, capacity planning, financial planning and overall risk management. Not to be used as a straitjacket, but to gain and maintain insight into progress, learning points, success factors and any obstacles (to be solved).

7. **Take a situational approach**
   There is no "one-size fits all" approach to implementing digital change. It depends on quite a few characteristics. Are the goals and results to be achieved crystal clear? Is it or is it not difficult to define the requirements and wishes for the solution SMARTly in advance? Do you expect many changes in the assignment and scope? How quickly are workable (intermediate) results needed? Or does accuracy and correctness take precedence over speed? Will a standard solution from the market be used? The answers to these questions determine the approach that best suits the situation. Sometimes it is best to work with an agile/ scrum approach, other times with a more classic waterfall approach. Or with a hybrid model.

8. **It is also behavioral change!**
   Digital transformation is also primarily about changing people's behavior. Many employees in your organization will undoubtedly find it difficult to change. It is important to get them to think in terms of opportunities instead of seeing barriers. To think and act from the outside in, instead of from professional knowledge and product specialization. And they will undoubtedly find it difficult to let go of the current situation. Company culture and risk-aversion are hugely important factors that can lead to procrastination and to the delay or failure of innovation and digitalization. Behavioral change is a difficult and important element of digital transformation. Make it explicit what digital transformation can mean for the people in your organization, what opportunities exist but also what will change. Let them think about the consequences for their own role and function, for the department they work in, for the organization as a whole and for the customers they deliver to. Pay particular attention to increasing your organization's willingness and ability to change, formulate actions, appoint change managers if necessary and monitor the effects.

Design and development

In the change management context, a distinction is often made between design and development. Change processes can be approached in two ways: the design approach and the development approach. This touches upon the field of tension between designing the desired situation and transforming the organization (in this case the people in the organization).

The design approach (solution-oriented) assumes a makeable future, craftsmanship, and attaches great significance to the exact models and designs *(engineering)* as formulated in the digital strategy and their predictive value. The aim is to achieve completeness, manufacturability and the delivery of blueprints. The methods and techniques used for this purpose have an instrumental, technical character.

The development approach (change-oriented) emphasizes the autonomous development and learning (*empowerment*) of organizations and people. The ability to plan this development is limited. The primary focus is on the process of realization and the involvement and motivation of the people in the organization. This is characterized by the organic or growth approach. The applied work methods concentrate primarily on guiding the natural development and creativity of the employees.

The starting point is that both approaches have validity. They will have to complement and balance each other. The design approach leads to the substantive digital strategy, the development approach brings the strategy to life in the organization.

9.4 FINALLY: AND NOW?

In this book, we started with the challenges your organization faces. Fundamental changes in customer needs and behavior. Newcomers to the market with new business models, who are running off with your customers. Technological advances that come along at a dizzying speed. But at the same time there is an inhibiting legacy and too much complexity in your organization that is blocking progress. There is a desperate need to change, to renew, to take control of digital transformation. The change agenda must be pushed, you cannot afford to sit back. This book provides guidance on how to tackle this.

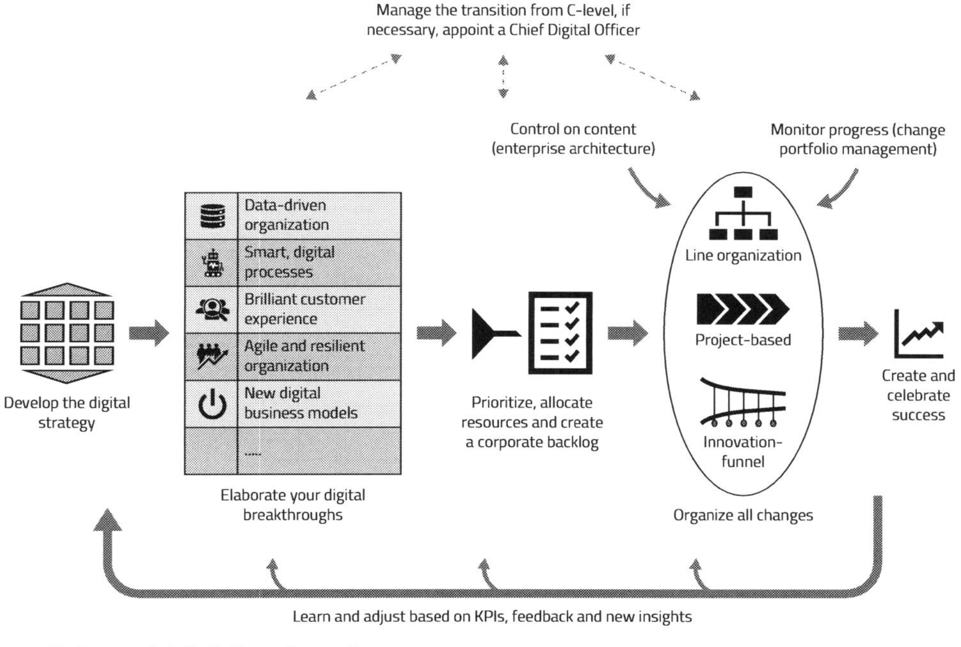

Figure 32 Successful digital transformation

Start by developing the digital strategy: a strategy that defines how digitization and the use of data can improve products and services, customer service and business processes. It guides the development and adoption of new technologies and even new business models.

Look from the outside in: what are the developments you are facing and what should you do with them? Which positions in the chain provide the most value? Which networks and partnerships are interesting? Which PMCs are you going to optimize and exploit, which are growth businesses, and which are you going to abandon? Which new digital business models do you see for the future? And for all these questions: what possibilities and opportunities does digital technology offer? Also look from the inside out: where are you now, what stage of digital transformation is your organization currently in, what digital capabilities do you possess, what about change capability, where do you need to improve and renew?

The five themes - data-driven organization, smart digital processes, brilliant customer experience, agile and resilient organization, and new digital business models - will undoubtedly be at the heart of your digital strategy. There are breakthroughs to be made on all five themes in the transition to the new world. Elaborate on these themes, focusing on your organization. Always start with the WHY - why is this so important? Set goals, see where you are now and define the situation to be reached, including the necessary digital capabilities you need to accelerate to phase 3. Formulate the desired change agenda for each theme.

Because not everything can be done at once, you should set priorities. Make a consideration based on value versus complexity and divide all change initiatives over the buckets "maintenance", "growth" and "innovation". This creates a "corporate backlog": a prioritized list of goals, actions and products that your organization wants to deliver or implement. Not a static one but one that is constantly being adjusted based on the latest insights.

Organize the changes in your organization:

- **In the line organization:** Place change initiatives here that belong to the primary responsibility of a value team, department or business unit, which do not require separate direction and additional resources.
- **Project-based**: Organize change initiatives that are project-based when the change task is one-time and complex, there are many stakeholders and additional resources are needed. Choose an appropriate approach: the classic waterfall method, agile/scrum or a hybrid model.
- **In an innovation tunnel:** House innovations here: ideas, initial thoughts, concepts that are worked out and tested on a small scale before scaling up. Screen these initiatives - until they require structural resources - and give them room to experiment.

Constantly keep the characteristics of the agile and resilient organization in mind when directing and implementing all changes:

- Learning by doing. Start small, work with living labs and pilots, constantly test whether you are on the right track. Take the next step based on insights from practice.
- Put the mandate where the expertise is, with those who are best positioned and most knowledgeable to make the right choices.
- Minimize dependencies and coordination; organize accountability as much as possible in the line organization.
- Provide a leadership style that fits the new way of working. Give space and trust to employees, facilitate and connect, set a good example.

Organize "horizontally" on all changes, with the role of enterprise architect assuming primary responsibility for steering on content. The enterprise architect develops and maintains the reference architecture, establishes (design) guidelines and standards, and facilitates all change initiatives in connecting to the reference architecture. A second steering role is that of the change portfolio manager. He takes care of overall planning and control of change initiatives in terms of coherence, time, capacity and money. He monitors the progress, guards dependencies and manages risks.

Preferably do not appoint a separate steering committee or change board, but steer the transition from the line, with full commitment and involvement of all C-level managers. In certain circumstances you may wish to consider appointing a (temporary) Chief Digital Officer (CDO) to take control of the transition. Continuously check whether you comply with the core principles for successful change. Monitor and evaluate continuously, making any adjustments based on the measurement of KPIs, feedback and new insights. "*When you can't change the direction of the wind, adjust your sails*" (H. Jackson Brown).

Literature list

Baghai, Mehrdad, Steve Coley, David White, Stephen Coley, *The Alchemy of Growth; Practical Insights for Building the Enduring Enterprise Paperback*, 2000

Bayens, Guido and Hans Tönissen, *Business architecture based on the Novius method; Working on a coherent business design*, Zaltbommel, 2013

Bayens, Guido, *Managers, Companies, Architecture; Successful entrepreneurship with enterprise architecture*, 2009

Belleghem, Steven van, *When digital becomes human; customer relationships in transformation*, 2014

Beijen, Marc, Eric Broos and Etienne Lucas, *Strategic deployment of ICT; a guide for business information management*, 2001

Beijen, Marc, Eric Broos and Etienne Lucas, *Business information planning; getting a grip on organizational change and ICT deployment*, 2003

Boonstra, Jaap, *Change management in 28 lessons*, 2013

Bossert, Oliver and Driek Desmet, *The platform play: How to operate like a tech company*, 2019

Boutetière, Hortense de la, Alberto Montagner en Angelika Reich, *Unlocking success in digital transformations*, 2018

Brackett, Michael H., Susan Earley and Mark Mosley, *The DAMA Guide to the Data Management Body of Knowledge*, 2012

Caudron, Jo and Dado van Peteghem, *Digital Transformation; preparing your organization for the future*, 2018

Denning, Stephen, *The Age of Agile: How Smart Companies Are Transforming the Way Work Gets Done*, 2017

Deuten, Jasper, *The power of platforms; new strategies for innovating in a digitizing world*, 2014

Gils, Bas van, *Data Management: a gentle introduction; Balancing theory and practice*, 2020

Hinssen, Peter, *Business/IT Fusion; how to move beyond Alignment and transform IT in your organization*, 2011

Kaplan, Robert and David Norton, *The execution premium*, 2008

Kotler, Philip and Gary Armstrong, *Principles of marketing - 17th edition*, 2017

Martin, Roger L., *The Design of Business: Why Design Thinking is the Next Competitive Advantage*, 2009

Moore, Michele, *Organizations struggle to make progress with their digital transformation investments*, 2018

Ohno, Taiichi, *Toyota Production System; beyond Large-Scale Production*, 1988

Osterwalder, Alexander and Yves Pigneur, *Business Model Generation; a handbook for visionaries, game changers and challengers*, 2010

Parker, Geoffrey G., *Platform Revolution; How Networked Markets Are Transforming the Economy and How to Make Them Work for You*, 2017

Ries, Eric, *The Lean Startup: How Today's Entrepreneurs Use Continuous Innovation to Create Radically Successful Businesses*, 2017

Sinek, Simon, *Start with why; How Great Leaders Inspire Everyone To Take Action*, 2011

Staffhorst, Sjoerd, Jeroen Stoop, Remco Bekker and Tjerk Hobma, *Business Transformation Framework; to get from strategy to execution*, 2016

Thomassen, Jean-Pierre and Eric de Haan, *Service Excellence; working towards excellent customer focus and exceptional customer experience with the Service Excellence Model*, 2016

Voskes, Tom and Alexandra Jankovich, *Make disruption work; a CEO handbook for digital transformation*, 2019

Westerman, George, Didier Bonnet and Andrew McAfee, *Leading Digital; Turning Technology into Business Transformation*, 2014

Zwiggelaar, Karin and Antoon van Luxemburg, *CIO 3.0; Making a difference in digital transformation*, 2016

Epilogue

Years ago, somewhere in the early days of my career, I wrote several books about strategy, information management and the use of information technology. I noticed how important it was for a consultancy firm to publish, not to keep knowledge, insights and experiences to yourself but to share them widely. A long time and many assignments later, I began to itch again; how nice it would be to capitalize on the consultancy experience I had built up over the years, to commit it to paper and to disseminate it. But it never happened, I simply did not have the time for it, one assignment after another came along and clients always took priority.

On March 15, 2020, we were told that the Netherlands was pretty much locked down by COVID-19. I was supposed to start a new assignment the following week, getting a stalled, international ERP implementation afloat. The following day I was told by the client that the job had been postponed indefinitely, understandably. All at once, unplanned, I had oceans of time in my schedule. A day later I crawled behind my laptop, three months later the first version of the manuscript was ready.

I could never have written this book without all my clients over the years. I have worked hard for them but also learned a great deal from them, and all that knowledge has ended up in this book in one way or another. I am particularly grateful to Marco Keim and Eric Sluis, for whom I was able to carry out many complex and challenging assignments. I am also grateful to Kees Smaling (especially for his insights on the agile organization, which we put into practice together), Casper van der Veen (for the basis of the phase model for digital transformation), Peter van Wijk (for his vision on customer-centric organizing) and Erik Hietkamp (for his deliciously astute review of the manuscript).

And of course, the many Novius colleagues, who provided the necessary input, critical notes and good suggestions: thank you all. I hope this book will stimulate and support you in helping clients and in the further growth of our consulting practice, of which we are so proud. Finally, my son Lars; he suggested (during our vacation in Spain) several more excellent improvements, muchos gracias!

Marc Beijen
marc.beijen@rhdhv.com

About the author

Marc Beijen (1963) has been the founder and partner of consultancy firm Novius since 1992. He studied Management Science and Industrial Engineering at the Technical University of Eindhoven. His vision on business transformation and strategic use of IT is based on over thirty years of experience in organizations, where strategic changes are high on the agenda. His personal (professional) mission is to help organizations improve performance through new organizational concepts in relation to the smart use of information technology. This book is therefore a "capitalization" of his consulting experience and thus not only bridges the gap between strategy and execution but also between theory and practice.

Marc Beijen is the founder of the successful "Business Transformation Framework", which is now used by many organizations in business and government as a standard for the translation of the business strategy to the tactical change calendar. The framework also forms the basis for the Novius Architecture Method. He is the author of several books and publications in the field of strategy, business information planning and information management.

You will find Marc Beijen in situations where innovation is taking place. Not as a goal but always to make things better for the customer, the employee, the shareholder or for society. This can happen in all sorts of situations but always when there are high ambitions and a lot of complexity to be broken down. He translates strategies and concepts into execution. He makes complexity manageable again, thinks several steps ahead, connects people and steers them directly in the right direction. He has implemented new product formulas at our national railroad company, supervised mergers, translated the strategies of financial organizations into execution (and helped realize them), supported clients with customer-oriented digitization, helped transform a bank into a digital platform, redesigned IT organizations, and so on.

Always with just one goal: substantial improvement through innovation

Index

-

3D printing 49
360° customer view 58, 62

A

(advanced) analytics 34
Agile 72
agile and resilient organization 15, 65
agile platform organization 82
algorithms 47
API 6
architecture 87
Artificial intelligence (AI) 6, 48

B

behavioral change 94
benefits tracking 89
big data 31
bitcoin 49
blockchain technology 6, 48
BPM (business process management) 16
brilliant customer experience 14, 53
business and IT alignment 10
business capability 54
business intelligence 34
business process management (BPM) 47
business rule engine 47
business rules 47
business strategy 20
Business-to-Business market 56
Business-to-Consumer market 56
Business Transformation Framework (BTF) 23

C

case management, (dynamic) 47
change agenda 32
change cycle 87
change resources 91
Chief Digital Officer 73
competencies 61, 82
competent employees and managers 15
complexity 5
continuous improvement process 89
control and compliance 18
cross-channel 58
crypto money 48
customer analytics 7
customer contact centers 54
customer data 58
Customer effort score (CES) 60
customer engagement 61
customer equity (CE) 60
customer experience (CX) 56
customer interaction 55
customer journey 11, 57
Customer lifetime value (CLV) 60
customer loyalty 54
customer need 24
customer orientation 10
customer process 60

Customer referral value (CRV) 60
customer satisfaction 60
cybercrime 7

D
data 7, 31
data centric 83
data-driven organization 15, 34
data governance 28
data lab 38
data management 32, 36, 37
data quality 37, 42
data security 37
data usage 36
descriptive analytics 35
Design Thinking 51, 71
DevOps 71
digital breakthrough 16
digital business models 75
digital disruption 4
digital DNA 18
digital fitness scan 21
digital innovation 2
digital platform development process 84
digital platforms 75
digital roadmap 23
digital strategy 3, 19, 87, 88
digital transformation 3
digital twin 48
digital twin of the organization (DTO) 48
digital vortex 4
digitization 2
disruptive business models 3

E
effective distribution 18
efficient processes 14
employee competencies 54
employee satisfaction 41
enterprise architecture (EA) 22
ethical requirements 6
excellent customer experience 10

F
first time right 60

G
GDPR (General Data Protection
Regulation) 34
guilds 69

H
Heraclitus 2
human augmentation 49

I
information and applications 25
innovation funnel 88
innovation lab 38
integrated customer view 59
internal entrepreneurship 15
internet of things (IoT) 6, 49
ISO 900x 42
IT4IT 66
IT control 71
IT delivery 66
IT function 67
IT projects 66
IT security 18

K
Kanban 71
Key Performance Indicators (KPIs) 50, 60

L
Lean 42, 71
Lean Startup 71, 79
legacy problems 2
legislation, national and international 7
life events 56
low-code software 65

M
make, buy or ally 18
management information 43
marketing Cs (Kotler) 24

master data management (MDM) 62
minimal viable product (MVP) 81
multichannel 58
multidisciplinary value teams 70

N

Net promoter score (NPS) 60
new markets and business models 5
next best action 58
next best offer 58
no-code software 65

O

omnichannel 58
omnichannel service 61
open IT architecture 83
open IT platform 16
opportunities analysis 21

P

personalized propositions 14
phase model for digital transformation 9
physical data 34
platform business 78
predictive analytics 35
prescriptive analytics 35
privacy 6
proactive service 61
process automation 47
process 24
process management 42
process management software 47
process mining 47
product focus 9
project management 71
pull strategy 19
push strategy 19

Q

quality management 37

R

relative complexity 91
relative value 91
robotic process automation (RPA) 47
robotization 7

S

SAFe (Scaled Agile Framework) 71
Scrum 71
shadow IT 67
Six Sigma 42
smart, digital processes 41
SMART (Specific, Measurable, Acceptable, Realistic and Time-bound) 20
social technologies 47
sourcing 18
Straight through processing (STP) 46
strategy-to-execution (S2E) 87
structured data 34

T

technology infrastructure 25
The Open Group 66
Three horizons of growth framework (McKinsey) 9
touchpoints 58

U

unique selling point (USP) 81
unstructured data 34

V

value/complexity matrix 90
value proposition 24

W

wearables 49
WFM (workflow management) 16
workflow management 47

Printed in Great Britain
by Amazon

67599345R10072